Texts and Monographs in Symbolic Computation

A Series of the Research Institute
for Symbolic Computation,
Johannes Kepler University Linz, Austria

Series Editors

Robert Corless; University of Western Ontario, Canada

Hoon Hong; North Carolina State University, USA

Tetsuo Ida; University of Tsukuba, Japan

Martin Kreuzer; Universität Passau, Germany

Bruno Salvy; INRIA Rocquencourt, France

Dongming Wang; Université Pierre et Marie Curie – CNRS, France

Peter Paule; Johannes Kepler University Linz, Austria

For further volumes:
http://www.springer.com/series/3073

Manuel Kauers · Peter Paule

The Concrete Tetrahedron

Symbolic Sums, Recurrence Equations,
Generating Functions, Asymptotic Estimates

SpringerWienNewYork

Manuel Kauers
Research Institute
for Symbolic Computation (RISC)
Johannes Kepler University Linz
4040 Linz, Austria
manuel.kauers@risc.jku.at

Peter Paule
Research Institute
for Symbolic Computation (RISC)
Johannes Kepler University Linz
4040 Linz, Austria
peter.paule@risc.jku.at

© Springer-Verlag/Wien 2011

SpringerWienNewYork is part of Springer Science+Business Media
springer.at

Typesetting: le-tex publishing services GmbH, Leipzig, Germany

Printed on acid-free paper
SPIN: 80023935

With 20 Figures

Library of Congress Control Number: 2010939462

ISSN 0943-853X
ISBN 978-3-7091-0444-6 SpringerWienNewYork

Preface

There are problems in mathematics which are so hard that they remain open for centuries. But many problems are not of this type. Often, being able to solve a problem just depends on knowing the right technique. This book is a book on techniques. More precisely, it is a book on techniques for solving problems about infinite sequences. Some of these techniques have belonged already to the repertoire of Euler and Gauss, others have been invented only a couple of years ago and are best suited for being executed in a computer algebra system.

A course on such techniques has been taught by the second author for almost twenty years at the Johannes Kepler University in Linz, Austria. The material of this course has served as a starting point in writing the present book. Besides the teaching experience with that course, also our personal experience with applying, developing, and implementing algorithms for combinatorial sequences and special functions has greatly influenced the text. The techniques we present have often worked for us, and we are convinced that they will also work for our readers.

We have several goals. The first is to give an overview of some of the most important mathematical questions about infinite sequences which can be answered by the computer. It may seem at first glance that a user manual of a computer algebra system could serve the same purpose. But this is not quite so, because a proper interpretation of a computational result may well require some knowledge of the underlying theory, and sometimes a problem may be solvable only by some variation of a classical algorithm. Our second goal is therefore also to explain how the algorithms work and on what principles they are based. Our third goal, finally, is to describe also some techniques which are suitable for traditional paper and pencil reasoning rather than for modern computer calculations. This is still useful because paper and pencil arguments are often needed to bring the problem at hand into an equivalent form which then can be handled by a computer.

We have included more than one hundred problems on which the techniques explained in the text can be tested. Some of the problems are meant to be solved on paper, others by using computer algebra. The reader will also find problems which only after some prior hand calculation can be completed electronically. These ones may consume some more time and thought than the others, but they surely offer the best training effect, because these problems give the most accurate impression of how the various techniques are applied in real life. Depending on the individual interests of the reader and on the particular problem at hand, it may be instructive not to resort to the built-in commands of a computer algebra system for, say, evaluating a certain sum in closed form, but instead to execute the algorithms described in the text in a step-by-step manner, using the computer algebra system only for basic operations such as adding, multiplying, or factoring polynomials, or for solving linear systems. We leave the choice of appropriate tools to the reader. But in either case, we want to stress explicitly that the use of computers is not just allowed but also strongly encouraged – not only for solving the problems in this book, but in general.

In order to spread this message as far as possible and to reach not only those who already know, but also undergraduate mathematics or computer science students and researchers from other mathematical or even non-mathematical disciplines, we have preferred concrete examples over abstract theorems, and informal explanations over formal proofs wherever this seemed to be in the interest of readability. We have also strived for reducing the assumed background knowledge as much as possible. The only knowledge which we will assume is some familiarity with basic algebraic notions (rings, fields, etc.), with linear algebra in finite dimensional vector spaces, and with some complex analysis in one variable. But not more.

With this background it is already possible to cover quite some material, including some of the most spectacular discoveries in the area, which have been made in the last decade of the 20th century. It is not our goal, however, to reach subjects of ongoing research. Important topics such as summation with $\Pi\Sigma$ fields, special function inequalities, holonomic functions in several variables, or objects defined by nonlinear recurrence or differential equations are not addressed at all. But even with this restriction, we do believe that the tools presented in this book are useful.

Let us finally take the opportunity to thank Victor Moll, Veronika Pillwein, and Carsten Schneider for valuable comments and enlighting discussions. We especially want to express our gratitude to Michael Singer and his students for carefully going through a preliminary version of the book. Their remarks and suggestions have been of great help in fixing errors and resolving confusion.

Hagenberg, October 2010

Manuel Kauers
Peter Paule

Contents

Chapter 1
Introduction

This is the story about four mathematical concepts which play a great role in many different areas of mathematics. It is a story about symbolic sums, recurrence (difference) equations, generating functions, and asymptotic estimates. In this book, we will study their key features in isolation or in combination, their mastery by paper and pencil or by computer programs, and their application to problems in pure mathematics or to "real world problems". To begin with, we take a look at a particular "real world problem" and see how sums, recurrence equations, generating functions, and asymptotic estimates may naturally arise in such a context. After having introduced these four main characters of this book informally by seeing them in action, we will then prepare the stage for their detailed study.

1.1 Selection Sort and Quicksort

Suppose you want to write a computer program that sorts a given array of numbers. There are many ways to do this. Which one should you choose if you want your program to be fast? The speed of a program can be estimated by counting the number of operations it performs in dependence of the input size. For a sorting algorithm, it is thus natural to ask how many comparisons of two elements are needed in order to bring them into the right order.

The Selection Sort algorithm sorts a given array $(a_1, \ldots, a_n)$ of numbers as follows. First it goes through the entire list and determines the minimum. Say a_k is the minimum. Then it exchanges a_1 and a_k so that the minimum is at the right position. Then it goes through the list again, starting at the second position, and searches for the minimum of the remaining elements. It is exchanged with a_2 so that it is also at the right position, and so on. An example with eight numbers is shown in Fig. 1.1.

M. Kauers, P. Paule, *The Concrete Tetrahedron*
© Springer-Verlag/Wien 2011

It is easy to see that this algorithm is correct. How many comparisons did it perform? In the first iteration, there were seven comparisons needed for finding the minimum of eight numbers. In the second iteration, six comparisons for seven numbers, and so on. Finally, in the seventh iteration, one comparison was needed for finding the minimum of two numbers. Altogether, we needed

$$7+6+5+4+3+2+1 = 28$$

comparisons.

This is if eight numbers are to be sorted. If there are n numbers, then there are

$$\sum_{k=1}^{n-1} k = 1+2+3+\ldots+(k-1)$$

comparisons, and we are faced with our first symbolic sum. It is a very easy one indeed, famous for having been solved by nine-year-old Gauss in elementary school who just added the sum term by term to its reverse and divided by two in order to get the closed form $\frac{1}{2}n(n-1)$:

$$\oplus \begin{cases} 1 \quad + \quad 2 \quad + \quad 3 \quad +\ldots+ (n-1) \quad = \sum_{k=1}^{n-1} k \\ (n-1) + (n-2) + (n-3) +\ldots+ \quad 1 \quad = \sum_{k=1}^{n-1} k \\ \hline n \quad + \quad n \quad + \quad n \quad +\ldots+ \quad n \quad = 2\sum_{k=1}^{n-1} k \end{cases}$$

$$\underbrace{}_{(n-1) \text{ summands}}$$

Or with black and white pebbles, as the ancient Greeks did it:

$$n(n-1) = 2\sum_{k=1}^{n-1} k$$

(In fact, the word "calculation" originates from the Greek word for "pebble".)

Either way, at this point the analysis of Selection Sort can be considered as settled. We have a closed form expression which tells us for every input size n how many comparisons the algorithm will perform to get n numbers sorted.

Let us now have a look at a second sorting algorithm. It is called Quicksort and was invented by Hoare in the 1960s in the course of first attempts to write programs for translating texts from one language into another. Our goal is to understand whether Quicksort deserves its name. Here is how the algorithm works. It first picks a random

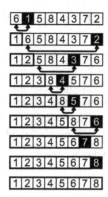

Fig. 1.1 Selection Sort

number from the list, called the pivot. Then it goes through the list and puts all the elements greater than the pivot to the right of it and all elements smaller than the pivot to the left of it. After that, the pivot is sitting at the correct position. The algorithm then proceeds recursively and sorts the part left of the pivot and the part right of the pivot. On our previous example, this may happen as shown in Fig. 1.2.

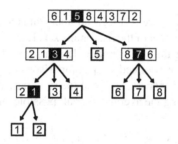

Fig. 1.2 Quicksort

We have chosen first 5 as pivot and brought the smaller numbers 2, 1, 3, and 4 to the left and the greater numbers 8, 7, 6 to the right. Then the algorithm is applied to $2, 1, 3, 4$. We have chosen 3 as pivot and found that no switch is needed. Then 2, 1 gets sorted to 1, 2, and the part to the left of 3 is done. The part to the right of 3 is trivial, as it consists of 4 only. Now we have $1, 2, 3, 4$ to the left of 5 and turn to the right. We have chosen 7 as pivot and exchanged 6 and 8 to bring them on the correct sides. The numbers 6 and 8 by themselves are trivially sorted, so $7, 8, 9$ are in order. This completes the execution.

How many comparisons did we perform? Let's count: in the first iteration we compared seven numbers to 5 for deciding to which side they belong. Then we compared three numbers to 3, one number to 1, and finally two numbers to 7. So there were

$$7 + 3 + 1 + 2 = 13$$

comparisons in total. This is clearly better than the 28 comparisons we needed with Selection Sort.

But sorting eight numbers with Quicksort does not always require 12 comparisons. The number of comparisons depends on the choice of the pivot. If, for example, it happens that the chosen pivot is always the minimum element of the (sub)list under consideration, then Quicksort just emulates Selection Sort and requires 28 comparisons for sorting eight numbers. In order to judge the performance of Quicksort, we need to know how many comparisons it needs "on average".

For $n \geq 0$, let c_n be that number. Then we have:

- $c_0 = 0$, because there is nothing to be sorted here.
- $c_1 = 0$, because every singleton list is sorted.
- $c_2 = 1$, because two elements can be sorted by a single comparison.
- $c_3 = \frac{8}{3}$: the pivot may be the minimum, the middle element or the maximum. Two comparisons are needed to determine this. If it is the minimum or the maximum, then the other two numbers are on the same side and it requires $c_2 = 1$ comparisons to sort them. If the pivot is the middle element, then there is one number to its left and one number to its right and no further comparison is needed. Assuming that each case is equally likely, we expect $2 + (1 + 0 + 1)/3$ comparisons on average.

In the general case, when we are sorting n numbers and choose a pivot p, that pivot can be the k-th smallest element of the list for any $k = 1, \ldots, n$. In every case, we need $n - 1$ comparisons to bring the $k - 1$ smaller elements to its left and the $n - k$ greater elements to the right. Then we need c_{k-1} comparisons on average to sort the left part and c_{n-k} comparisons on average to sort the right part, thus $n - 1 + c_{k-1} + c_{n-k}$ comparisons in total. Taking the average over all possible choices for k, we find

$$c_n = \frac{1}{n} \sum_{k=1}^{n} \left((n-1) + c_{k-1} + c_{n-k} \right) = (n-1) + \frac{1}{n} \sum_{k=1}^{n} (c_{k-1} + c_{n-k}) \qquad (n \geq 1).$$

So we now know what c_n is. In some sense. The answer may not be satisfactory because it is not in "closed form", but at least we can use it for computing c_n recursively for large n. (Observe that this is possible because all the c_{k-1} and c_{n-k} appearing on the right-hand side have smaller index than n.) We find

n	0	1	2	3	4	5	6	7	8	9	10
c_n	0	0	1	$\frac{8}{3}$	$\frac{29}{6}$	$\frac{37}{5}$	$\frac{103}{10}$	$\frac{472}{35}$	$\frac{2369}{140}$	$\frac{2593}{126}$	$\frac{30791}{1260}$

and it would be easy to provide more numbers. So we can now decide for every fixed and specific n whether sorting n numbers is better done by Quicksort or by Selection Sort. Indeed, we have $c_n < \frac{1}{2}n(n-1)$ for $n = 3, \ldots, 10$ and Fig. 1.3 suggests that this is so for all large n, and that the advantage of Quicksort over Selection Sort gets even much better as n grows. But a picture is not a proof.

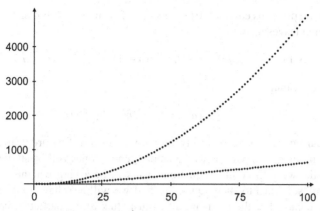

Fig. 1.3 Comparison of c_n (diamonds) to $\frac{1}{2}n(n-1)$ (circles)

1.2 Recurrence Equations

For a rigorous argument, we need a better representation of the sequence $(c_n)_{n=0}^{\infty}$. Some cleanup is straightforward:

$$c_n = (n-1) + \frac{1}{n}\sum_{k=1}^{n}(c_{k-1}+c_{n-k}) \qquad \text{(split the sum into two)}$$

$$= (n-1) + \frac{1}{n}\left(\sum_{k=1}^{n}c_{k-1}+\sum_{k=1}^{n}c_{n-k}\right) \qquad \text{(reverse the order of summation)}$$

$$= (n-1) + \frac{1}{n}\left(\sum_{k=1}^{n}c_{k-1}+\sum_{k=1}^{n}c_{n-(n-k+1)}\right) \qquad \text{(now it is twice the same sum)}$$

$$= (n-1) + \frac{2}{n}\sum_{k=1}^{n}c_{k-1} \qquad \text{(let the sum start from } k=0\text{)}$$

$$= (n-1) + \frac{2}{n}\sum_{k=0}^{n-1}c_k.$$

This looks just slightly better. The unpleasant thing is that c_k appears in a sum expression. To get rid of this disturbing term, we apply an adapted version of Gauss's summation trick. First multiply the equation by n to obtain the form

$$nc_n = n(n-1) + 2\sum_{k=0}^{n-1}c_k \qquad (n \geq 1).$$

Replace n by $n+1$ to obtain

$$(n+1)c_{n+1} = (n+1)n + 2\sum_{k=0}^{n}c_k \qquad (n \geq 0),$$

and subtract the previous equation from this one. This makes all but one term of the sum cancel and leaves us with

$$(n+1)c_{n+1} - nc_n = (n+1)n - n(n-1) + 2c_n \qquad (n \geq 0),$$

or, after some cleanup,

$$(n+1)c_{n+1} - (n+2)c_n = 2n \qquad (n \geq 0).$$

This is a recurrence, or more precisely, an inhomogeneous first order linear difference equation. Such equations as well as more general ones will be studied later in this book. Like with differential equations, we will be asking whether there exist "closed form" solutions to such equations, and we will also have to discuss what this actually means. For the moment, let us just stick to the particular equation at hand. What can we do to solve it?

Discard first, to make things simpler, the term $2n$ on the right-hand side, and consider just the equation

$$(n+1)h_{n+1} - (n+2)h_n = 0 \qquad (n \geq 0).$$

This is called the associated homogeneous equation. (We write here h_n for the unknown sequence instead of c_n, because the equation is not true for the sequence c_n we defined earlier.) If we rewrite this equation in the form

$$h_{n+1} = \frac{n+2}{n+1} h_n,$$

we can apply it repeatedly to itself:

$$
\begin{aligned}
h_{n+1} &= \frac{n+2}{n+1} h_n \\
&= \frac{n+2}{n+1} \frac{n+1}{n} h_{n-1} \\
&= \frac{n+2}{n+1} \frac{n+1}{n} \frac{n}{n-1} h_{n-2} \\
&= \frac{n+2}{n+1} \frac{n+1}{n} \frac{n}{n-1} \frac{n-1}{n-2} h_{n-3} \\
&= \cdots
\end{aligned}
$$

Continuing this process all the way down to h_0 and performing the obvious cancellations, we find $h_{n+1} = (n+2)h_0$, or

$$h_n = (n+1)h_0 \qquad (n \geq 0).$$

This is the general solution of the homogeneous equation; the constant h_0 is up to our choice.

The solution of the homogeneous equation suggests what the solution of the inhomogeneous equation might look like. We make an ansatz $c_n = (n+1)u_n$ for some unknown sequence $(u_n)_{n=0}^{\infty}$ in place of the undetermined constant h_0 in the solution of the homogeneous equation. Like for differential equations, this approach is known as *variation of constants*. Going with $c_n = (n+1)u_n$ into the inhomogeneous equation

$$(n+1)c_{n+1} - (n+2)c_n = 2n$$

brings us to

$$(n+1)(n+2)u_{n+1} - (n+2)(n+1)u_n = 2n,$$

which has the nice feature that the coefficients in front of u_{n+1} and u_n are identical. So we can divide them to the other side and get

$$u_{n+1} - u_n = \frac{2n}{(n+1)(n+2)}.$$

Summing the equation on both sides gives

$$\sum_{k=0}^{n-1}(u_{k+1} - u_k) = \sum_{k=0}^{n-1}\frac{2k}{(k+1)(k+2)}.$$

As all terms in the sum on the left-hand side cancel, except for u_n and u_0 (Problem 1.1), we obtain

$$u_n = u_0 + \sum_{k=0}^{n-1}\frac{2k}{(k+1)(k+2)}.$$

Because of $0 = c_0 = (0+1)u_0$, we have $u_0 = 0$ and therefore

$$c_n = 2(n+1)\sum_{k=0}^{n-1}\frac{k}{(k+1)(k+2)} \qquad (n \geq 1)$$

We have solved the recurrence for c_n in terms of a sum.

1.3 Symbolic Sums

Can this sum be evaluated in closed form? The answer depends on what expressions we accept as closed form, but as we will see in Chap. 5, there is no closed form for this sum in the usual sense of the word. Very roughly speaking, this means that we cannot get rid of the summation sign here. But there are many different ways of writing c_n with a summation sign, the formula above is just one of them. We can obtain an alternative representation, which in some sense is nicer, by breaking the summand into partial fractions. We have

$$\frac{k}{(k+1)(k+2)} = \frac{2}{k+2} - \frac{1}{k+1} = \frac{1}{k+2} + \left(\frac{1}{k+2} - \frac{1}{k+1}\right) \qquad (k \geq 0).$$

If we sum this equation for k from 0 to $n-1$, then the part in the brackets telescopes and we obtain

$$\sum_{k=0}^{n-1} \frac{k}{(k+1)(k+2)} = \sum_{k=0}^{n-1} \left(\frac{1}{k+2} + \left(\frac{1}{k+2} - \frac{1}{k+1} \right) \right)$$

$$= \sum_{k=0}^{n-1} \frac{1}{k+2} + \frac{1}{n+1} - 1$$

$$= \sum_{k=2}^{n+1} \frac{1}{k} + \frac{1}{n+1} - 1$$

$$= \sum_{k=1}^{n} \frac{1}{k} + \frac{1}{n+1} - 1 + \frac{1}{n+1} - 1$$

$$= \sum_{k=1}^{n} \frac{1}{k} - \frac{2n}{n+1} \qquad (n \geq 1).$$

It follows that

$$c_n = 2(n+1) \sum_{k=1}^{n} \frac{1}{k} - 4n \qquad (n \geq 1).$$

The numbers

$$H_n := \sum_{k=1}^{n} \frac{1}{k} \quad (n \geq 1), \qquad H_0 := 0$$

are known as the *harmonic numbers,* and we will study them more closely in Sect. 7.1. With this notation, we can formulate our result as follows: If c_n denotes the average number of comparisons performed by applying Quicksort to a list of n numbers, then

$$c_n = 2(n+1)H_n - 4n \qquad (n \geq 1).$$

1.4 Generating Functions

A generating function is an alternative representation of a sequence, to which operations can be applied that resemble operations from calculus, like integration or differentiation. The generating function of a sequence $(a_n)_{n=0}^{\infty}$ is simply defined as the power series

$$a(x) := \sum_{n=0}^{\infty} a_n x^n.$$

We will later (Chap. 2) insist that our power series be *formal* power series rather than analytic functions, so that we are dispensed from possible convergence or divergence issues that may arise. But for the moment, let us suppress these technical details.

As an example, consider the sequence $(a_n)_{n=0}^\infty$ with $a_n = 1$ for all n. Its generating function is

$$\sum_{n=0}^\infty x^n = \frac{1}{1-x}.$$

The expression $1/(1-x)$ encodes the whole infinite sequence $(a_n)_{n=0}^\infty$ in finite terms. Incidentally, our sequence has the simple closed form "1", which also encodes the whole infinite sequence in finite terms, but in more complicated situations it can happen that there is no closed form for the terms of a sequence, but a closed form for its generating function, or vice versa.

For example, we have noted (and will prove later) that the harmonic numbers $H_n = \sum_{k=1}^n \frac{1}{k}$ have no closed form. What about their generating function? Let

$$H(x) := \sum_{n=1}^\infty H_n x^n.$$

We must derive information about $H(x)$ from the available information about the coefficient sequence H_n. Knowing that

$$H_{n+1} = \sum_{k=1}^{n+1} \frac{1}{k} = \frac{1}{n+1} + \sum_{k=1}^n \frac{1}{k} = \frac{1}{n+1} + H_n \qquad (n \geq 0),$$

we find (recall that we had defined $H_0 = 0$)

$$H(x) = \sum_{n=1}^\infty H_n x^n = \sum_{n=0}^\infty H_{n+1} x^{n+1} = \sum_{n=0}^\infty \left(\frac{1}{n+1} + H_n \right) x^{n+1}$$

$$= \sum_{n=0}^\infty \frac{1}{n+1} x^{n+1} + \sum_{n=0}^\infty H_n x^{n+1}.$$

The second term on the right is recognized as $xH(x)$, and about the first we can reason (somewhat sloppily) that

$$\sum_{n=0}^\infty \frac{1}{n+1} x^{n+1} = \sum_{n=0}^\infty \left(\int_0^x t^n \, dt \right) = \int_0^x \left(\sum_{n=0}^\infty t^n \right) dt = \int_0^x \frac{1}{1-t} dt = -\log(1-x).$$

Putting this together with the previous calculation, we obtain

$$H(x) = -\log(1-x) + xH(x),$$

which implies

$$H(x) = -\frac{\log(1-x)}{1-x} = \frac{1}{1-x} \log \frac{1}{1-x}.$$

So the generating function $H(x)$ of the sequence of harmonic numbers admits a closed form even though the harmonic numbers themselves do not.

Now consider again the sequence c_n that counts the average number of comparisons needed by Quicksort. Recall that we had derived the recurrence equation

$$(n+1)c_{n+1} = (n+1)n + 2\sum_{k=0}^{n} c_n \qquad (n \geq 0). \tag{R}$$

Let

$$c(x) := \sum_{n=0}^{\infty} c_n x^n$$

be the generating function of c_n. Observe that then

$$c'(x) = \sum_{n=1}^{\infty} c_n n x^{n-1} = \sum_{n=0}^{\infty} (n+1)c_{n+1} x^n$$

is the generating function of the left hand side of the equation. For the right hand side, we need $\sum_{n=0}^{\infty} n(n+1)x^n$. We have

$$\frac{1}{1-x} = \sum_{n=0}^{\infty} x^n$$

$$\implies \quad \frac{d}{dx}\frac{1}{1-x} = \sum_{n=1}^{\infty} nx^{n-1} = -\frac{1}{(1-x)^2}$$

$$\implies \quad \frac{d^2}{dx^2}\frac{1}{1-x} = \sum_{n=2}^{\infty} n(n-1)x^{n-2} = \sum_{n=1}^{\infty} (n+1)nx^{n-1} = \frac{1}{x}\sum_{n=0}^{\infty} n(n+1)x^n.$$

Therefore

$$\sum_{n=0}^{\infty} (n+1)nx^n = x\frac{d^2}{dx^2}\frac{1}{1-x} = \frac{2x}{(1-x)^3},$$

and we have found the generating function for the first term on the right hand side of (R). For the second, recall that the multiplication law for power series reads

$$\left(\sum_{n=0}^{\infty} a_n x^n\right)\left(\sum_{n=0}^{\infty} b_n x^n\right) = \sum_{n=0}^{\infty} \left(\sum_{k=0}^{n} a_k b_{n-k}\right)x^n.$$

Using this law from right to left with $a_n = c_n$ and $b_n = 1$, we find

$$\sum_{n=0}^{\infty} \left(\sum_{k=0}^{n} c_k\right)x^n = \left(\sum_{n=0}^{\infty} c_n x^n\right)\left(\sum_{n=0}^{\infty} x^n\right) = \frac{1}{1-x}c(x).$$

Putting things together, the recurrence for c_n translates into the differential equation

$$c'(x) = \frac{2x}{(1-x)^3} + \frac{2}{1-x}c(x)$$

for the generating function. The initial condition $c_0 = 0$ translates into the initial condition $c(0) = 0$. This differential equation can be solved in closed form, for instance with the help of a computer algebra system, the result being

$$c(x) = -\frac{2x}{(1-x)^2} - \frac{2}{(1-x)^2}\log(1-x).$$

(To be honest, this is not exactly the form that Maple or Mathematica return, see Problem 1.7.)

Let us transform the information about $c(x)$ into information about the coefficients c_n. We handle the two terms in the closed form of $c(x)$ separately. First,

$$-\frac{2x}{(1-x)^2} = (-2x)\sum_{n=0}^{\infty} nx^{n-1} = \sum_{n=0}^{\infty}(-2n)x^n,$$

as we had seen earlier. Secondly,

$$-\frac{2}{(1-x)^2}\log(1-x) = \frac{2}{1-x}H(x) = 2\left(\sum_{n=0}^{\infty}x^n\right)\left(\sum_{n=0}^{\infty}H_nx^n\right)$$
$$= \sum_{n=0}^{\infty}\left(2\sum_{k=0}^{n}H_k\right)x^n,$$

using the generating function for harmonic numbers found earlier. Putting the two together, we find

$$c(x) = \sum_{n=0}^{\infty} c_nx^n = \sum_{n=0}^{\infty}\left(-2n+2\sum_{k=0}^{n}H_k\right)x^n$$

and thus

$$c_n = -2n+2\sum_{k=0}^{n}H_k \qquad (n\geq 0).$$

This is not quite the same formula that we found before, but it is an equivalent one, because of the identity

$$\sum_{k=0}^{n}H_k = (n+1)H_n - n \qquad (n\geq 0).$$

Identities like this can nowadays be found by computer algebra packages. In Appendix A.5 we give some references to actual pieces of software for discovering, proving, and manipulating identities. Using the identity above, we are led to a form that matches our previous result about c_n. Alternatively, we can do the coefficient extraction by observing that

$$H'(x) = -\frac{1}{(1-x)^2}\log(1-x) + \frac{1}{(1-x)^2},$$

so

$$c(x) = -\frac{2x}{(1-x)^2} + 2H'(x) - \frac{2}{(1-x)^2}$$

$$= \sum_{n=0}^{\infty}(-2n)x^n + 2\sum_{n=1}^{\infty}nH_n x^{n-1} - 2\sum_{n=1}^{\infty}nx^{n-1}$$

$$= \sum_{n=0}^{\infty}(-2n)x^n + 2\sum_{n=0}^{\infty}(n+1)H_{n+1}x^n - 2\sum_{n=0}^{\infty}(n+1)x^n$$

$$= \sum_{n=0}^{\infty}\left(-2n + 2(n+1)H_{n+1} - 2(n+1)\right)x^n,$$

and

$$c_n = -2n + 2(n+1)H_{n+1} - 2(n+1) = -4n + 2(n+1)H_n \qquad (n \geq 0)$$

gives us back the formula obtained earlier without having to evaluate the sum over H_k.

1.5 Asymptotic Estimates

Of course there are plenty of sequences $(a_n)_{n=0}^{\infty}$ which do not admit a closed form representation and whose generating function $a(x)$ does not have a closed form either. It also happens that a sequence and/or its generating functions does have a closed form which is too complicated to be of any use. Even in such unfavorable situations it is often possible to find a simple description for the asymptotic behavior of the sequence $(a_n)_{n=0}^{\infty}$. We say that a sequence $(a_n)_{n=0}^{\infty}$ is asymptotically equivalent to a sequence $(b_n)_{n=0}^{\infty}$, written $a_n \sim b_n$ $(n \to \infty)$, if

$$\lim_{n\to\infty}\frac{a_n}{b_n} = 1.$$

For example, in order to describe the asymptotics of

$$c_n = 2(n+1)H_n - 4n \qquad (n \geq 1),$$

we can exploit known facts about harmonic numbers. Specifically, it was already shown by Euler that the limit

$$\gamma := \lim_{n\to\infty}(H_n - \log n)$$

exists and is finite. The approximate value is $\gamma \approx 0.577216$, but this is not really relevant for our purpose. All we need for the moment is the slightly weaker result that $H_n \sim \log(n)$ $(n \to \infty)$. With this information at hand, it is then an easy matter to deduce that

$$c_n \sim 2n\log n \qquad (n \to \infty).$$

At this point it is fairly clear that on average Quicksort is better than Selection Sort.

It is also possible to obtain asymptotic information from a generating function. To this end, one interprets the generating function as a complex analytic function in a suitable neighborhood of the origin and considers the position and the type of its singularities which are closest to the origin. The asymptotic behavior of the generating function near these singularities governs the asymptotic behavior of the series coefficients in the expansion at the origin. Similarly as for sequences, two complex functions a, b are said to be asymptotically equivalent at a point ζ, written $a(z) \sim b(z)$ $(z \to \zeta)$, if

$$\lim_{z \to \zeta} \frac{a(z)}{b(z)} = 1.$$

There are general theorems which link asymptotic equivalence of the generating functions at the singularities closest to the origin with the asymptotic equivalence of the coefficient sequences. For example, if $a(z) = \sum_{n=0}^{\infty} a_n z^n$ for z in some open neighborhood of zero containing 1, then

$$a(z) \sim (1-z)^{\alpha} \log(1-z)^{\beta} \ (z \to 1) \iff a_n \sim \frac{(-1)^{\beta}}{\Gamma(-\alpha)} n^{-\alpha-1} \log(n)^{\beta} \ (n \to \infty)$$

for any $\alpha \notin \mathbb{N}$ and $\beta \in \mathbb{N}$ (see Sect. IV.2 of [21]). Applying this result to the generating function

$$c(x) = -\frac{2x}{(1-x)^2} - \frac{2}{(1-x)^2} \log(1-x)$$

gives us back what we know already:

$$c_n \sim 2n \log n \qquad (n \to \infty).$$

1.6 The Concrete Tetrahedron

In the preceding sections we have seen sums, recurrence equations, generating functions as well as asymptotic estimates in action. We have seen that these concepts do not stand for themselves, but that there are close connections among them: A recurrence equation has been deduced from a symbolic sum, a generating function was deduced from a recurrence, and asymptotic estimates can be deduced when the generating function is available. And this was just a glimpse on what can be done. There are many other connections. In fact, there are connections from each of the four concepts to every other. A pictorial description of the mutual connections is given by the tetrahedron in Fig. 1.4. We call it the *Concrete Tetrahedron*.

The attribute *concrete* is a reference to the book "Concrete Mathematics" by Graham, Knuth, and Patashnik [24], where a comprehensive introduction to the subject is provided. Following the authors of that book, we understand *concrete* not in contrast to *abstract,* but as a blend of the two words *con*-tinuous and dis-*crete,* for it

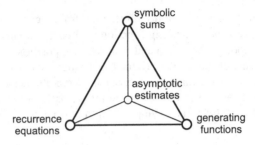

Fig. 1.4 The Concrete Tetrahedron

is both continuous and discrete mathematics that is applied when solving concrete problems. In most stages of our discourse, the book by Graham, Knuth, Patashnik can be used as a reference to many additional techniques, applications, stories, etc.

However, the present book is not meant to be merely a summary of "Concrete Mathematics". We have a new twist to add to the matter, and this is computer algebra. In the last decade of the 20th century, many algorithms have been discovered by which much of the most tedious and error-prone work about the four vertices of the Concrete Tetrahedron can be performed by simply pressing a button. We believe that a mathematics student of the 21st century must be able to use these algorithms, and so we will devote a great part of this book to explaining what can and should be left to a computer, and what can and should be still better done the traditional way.

Stated in most general terms, the techniques we present in this book apply to infinite sequences. However, as far as the computer is concerned, we must apply some restrictions. In order for a mathematical object to be stored faithfully on a computer, we must come up with some finite representation of it, say an "expression" or a defining equation. This necessarily excludes most sequences from the game, as there are uncountably many sequences but only countably many expressions. There can, for this reason, never be an algorithm that takes arbitrary infinite sequences as input and performs some operations on them. An algorithm can only operate on finite data and therefore only on sequences that are given in terms of some finite data.

For example, we may choose to restrict ourselves to sequences $(a_n)_{n=0}^{\infty}$ where $a_n = p(n)$ for some polynomial p, and we choose that polynomial p as the expression defining the sequence. The sequences which admit such a representation form a class, and a variety of algorithms is available for solving problems about the sequences living in this class. Other ways of representing sequences have been found in the past, they define larger classes of sequences, and algorithms are available for them as well. As a rule, the bigger a class is, the fewer questions about its elements can be answered by an algorithm.

Our plan for this book is to study the Concrete Tetrahedron for some of the most important classes of sequences. These are: polynomial sequences, C-finite sequences, hypergeometric sequences, sequences whose generating function is algebraic (alge-

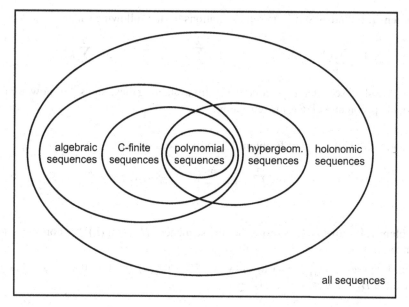

Fig. 1.5 The classes of sequences studied in this book

braic sequences), and holonomic sequences. Their mutual inclusions are depicted in Fig. 1.5.

As we make our tour through these classes, we will look at the Concrete Tetrahedron from different perspectives. We will see different techniques for dealing with sums, recurrences, generating functions, and asymptotics that are adequate in different contexts.

1.7 Problems

Problem 1.1 (Telescoping) Let $f\colon \mathbb{Z} \to \mathbb{C}$ and $a, b \in \mathbb{Z}$ with $a \le b$.

1. Show that
$$S(a,b) := \sum_{k=a}^{b} \big(f(k+1) - f(k)\big)$$
satisfies $S(a,b) = f(b+1) - f(a)$.

2. Assuming that $f(k) \ne 0$ for $a \le k \le b$, show that
$$P(a,b) := \prod_{k=a}^{b} \frac{f(k+1)}{f(k)}$$
satisfies $P(a,b) = f(b+1)/f(a)$.

Problem 1.2 Find closed form representations for the following sums:

$$1. \ \sum_{k=0}^{n} (2k+1), \qquad 2. \ \sum_{k=0}^{n} k^2, \qquad 3. \ \sum_{k=0}^{n} k^3.$$

(*Hint:* We shall see later a systematic way for evaluating these sums. For now, try to solve the problem with ad-hoc reasoning.)

Problem 1.3 Use part 2 of Problem 1.1 to find a closed form for

$$a_n := \prod_{k=2}^{n} \left(1 - \frac{1}{k^2}\right) \qquad (n \geq 2).$$

Problem 1.4 Prove Euler's result that the sequence $(H_n - \log(n))_{n=1}^{\infty}$ converges to a finite limit.

(*Hint:* Use $\log(n) = \int_1^n 1/x\,dx$ to show that $H_n - \log(n) \geq 0$ for all $n \geq 1$, and then check that $H_n - \log(n)$ is decreasing.)

Problem 1.5 Find the generating function for $(a_n)_{n=0}^{\infty}$ where

$$1. \ a_n = n, \qquad 2. \ a_n = n^2, \qquad 3. \ a_n = 1/(n+1).$$

Problem 1.6 Let $a(x)$ be the generating function of some infinite sequence $(a_n)_{n=0}^{\infty}$. What is the generating function for the sequence $(2^n a_n)_{n=0}^{\infty}$?

Problem 1.7 A contemporary computer algebra system returned

$$c(x) = \frac{-2(x + \log(x-1) - i\pi)}{(x-1)^2}$$

as the solution of the differential equation

$$c'(x) = \frac{2x}{(1-x)^3} + \frac{2}{1-x} c(x), \quad c(0) = 0.$$

Argue that this is equivalent to the form given in the text.

Chapter 2
Formal Power Series

This chapter is somewhat different from the other chapters in this book. It contains not many symbolic sums, hardly any recurrence equations, and only few asymptotic estimates. We provide here the algebraic background on which the notion of a generating function rests. The results developed here apply in full generality to all sequences, not just to some limited class of sequences as will be the case in all the following chapters.

2.1 Basic Facts and Definitions

An infinite sequence $(a_n)_{n=0}^\infty$ in a field $\mathbb{K}$ is simply a map from $\mathbb{N}$ to $\mathbb{K}$, assigning to each $n \in \mathbb{N}$ some element a_n from the field $\mathbb{K}$. The set of all sequences in $\mathbb{K}$, denoted by $\mathbb{K}^\mathbb{N}$, forms a vector space over $\mathbb{K}$ if we define addition and scalar multiplication termwise:

$$(a_n)_{n=0}^\infty + (b_n)_{n=0}^\infty := (a_n + b_n)_{n=0}^\infty \quad \text{and} \quad \alpha(a_n)_{n=0}^\infty := (\alpha a_n)_{n=0}^\infty \ (\alpha \in \mathbb{K}).$$

We can turn this vector space into a ring by defining a multiplication. There are several ways of doing so. One possible multiplication is the *Hadamard product*, which is also defined termwise:

$$(a_n)_{n=0}^\infty \odot (b_n)_{n=0}^\infty := (a_n b_n)_{n=0}^\infty.$$

It is immediate that $\mathbb{K}^\mathbb{N}$ equipped with addition and Hadamard product is a commutative ring, for it inherits the necessary laws (commutativity, associativity, distributivity, neutral elements) directly from $\mathbb{K}$.

However, $\mathbb{K}^\mathbb{N}$ is far from being a field, even if $\mathbb{K}$ is, because for $\mathbb{K}^\mathbb{N}$ to be a field, we would need a multiplicative inverse for every nonzero sequence, while, for example, the sequence

$$1, 0, 1, 0, 1, 0, 1, 0, 1, 0, 1, 0, \dots$$

M. Kauers, P. Paule, *The Concrete Tetrahedron*
© Springer-Verlag/Wien 2011

does not have a multiplicative inverse. In fact, a sequence $(a_n)_{n=0}^{\infty}$ has a multiplicative inverse if and only if $a_n \neq 0$ for all n. And even worse, there are zero divisors in the ring $\mathbb{K}^{\mathbb{N}}$: If we multiply the sequence above with the sequence

$$0, \ 1, \ 0, \ 1, \ 0, \ 1, \ 0, \ 1, \ 0, \ 1, \ 0, \ 1, \ldots$$

we get zero identically. Zero divisors in a ring are typically a sign that things may get hairy, so let us turn to a different way of defining multiplication for sequences. This second way is known as *Cauchy product* (or convolution) and is defined via

$$(a_n)_{n=0}^{\infty} \cdot (b_n)_{n=0}^{\infty} := (a_n)_{n=0}^{\infty}(b_n)_{n=0}^{\infty} := (c_n)_{n=0}^{\infty} \quad \text{where} \quad c_n = \sum_{k=0}^{n} a_k b_{n-k}.$$

Also this multiplication turns the vector space $\mathbb{K}^{\mathbb{N}}$ into a ring:

Theorem 2.1 $\mathbb{K}^{\mathbb{N}}$ *together with termwise addition and Cauchy product as multiplication forms a commutative ring.*

The proof is straightforward by checking the necessary laws (Problem 2.1). Because of the formal similarity of the Cauchy product with the multiplication law for power series, it proves convenient to also borrow the notation for power series for writing down sequences. Consequently, in addition to $(a_n)_{n=0}^{\infty}$, we shall use the notation

$$a(x) = \sum_{n=0}^{\infty} a_n x^n,$$

with x an indeterminate. When this notation is used, it is customary to speak of a *formal power series* instead of a "sequence", although both notions actually refer to the same object. It is also customary to write $(\mathbb{K}[[x]], +, \cdot)$, or $\mathbb{K}[[x]]$ for short, for the ring of all formal power series (i.e., sequences) with Cauchy product as multiplication, and to write $(\mathbb{K}^{\mathbb{N}}, +, \odot)$, or $\mathbb{K}^{\mathbb{N}}$ for short, for the ring of sequences with Hadamard product as multiplication.

We will call the formal power series $a(x)$ the *generating function* of the sequence $(a_n)_{n=0}^{\infty}$, and we will see that a lot of calculations can be performed on formal power series as if they were analytic functions. It must be emphasized however, that we do *not* intend to regard $a(x)$ as a function mapping points x from some domain to another. In particular, it is pointless to ask for the radius of convergence of a formal power series. Nevertheless, we will also freely use notation for common functions such as $\exp(x)$, $\log(x+1)$, $\sin(x)$, $\sqrt{1-x^2}$, etc., even when we actually mean the formal power series whose coefficients are as in the Taylor expansion of these analytic functions. Since most of our discourse will stay entirely inside the algebraic framework of formal power series, no confusion will arise from this.

We will next show that $\mathbb{K}[[x]]$ belongs to a much more well-behaved class of rings than $\mathbb{K}^{\mathbb{N}}$ does.

Theorem 2.2 $\mathbb{K}[[x]]$ *is an integral domain.*

Proof. We have to show that $a(x) \neq 0$ and $b(x) \neq 0$ implies $a(x)b(x) \neq 0$. So let $a(x) = \sum_{n=0}^{\infty} a_n x^n$, $b(x) = \sum_{n=0}^{\infty} b_n x^n$ be two non-zero formal power series. A formal power series is nonzero if and only if it has at least one nonzero coefficient. Therefore there exist minimal indices i, j such that $a_i \neq 0$ and $b_j \neq 0$, respectively. Now consider the product $c(x) = \sum_{n=0}^{\infty} c_n x^n := a(x)b(x)$. By definition,

$$c_n = \sum_{k=0}^{n} a_k b_{n-k} \quad (n \geq 0).$$

We have $a_k = 0$ for $k < i$ and $b_{n-k} = 0$ for $n - k < j$ or $k > n - j$. Therefore, for $n = i + j$ we find

$$c_{i+j} = \sum_{k=0}^{i+j} a_k b_{i+j-k} = a_i b_j.$$

Since both a_i and b_j are nonzero, so is c_{i+j}, and therefore $c(x)$ cannot be the zero series. $\square$

If $a(x) = \sum_{n=0}^{\infty} a_n x^n$ is a formal power series, we will use the notation $[x^k]a(x) := a_k$ to refer to the coefficient of x^k. For fixed $k \in \mathbb{N}$, coefficient extraction $[x^k] : \mathbb{K}[[x]] \to \mathbb{K}$ is a linear map. For $k = 0$, also the notations $a(x)|_{x=0} := a(0) := [x^0]a(x)$ are used. The coefficient of x^0 is called the *constant term* of $a(x)$.

The definition of the Cauchy product may seem somewhat unmotivated at first sight, but there is a natural combinatorial interpretation for it. A combinatorial class $\mathscr{A}$ is a set of objects with some natural number, called the size, associated to each object in $\mathscr{A}$. For instance, if $\mathscr{A}$ is some collection of graphs, the number of vertices may be taken as their size. If a_n denotes the number of objects in $\mathscr{A}$ of size n, then the power series $a(x) = \sum_{n=0}^{\infty} a_n x^n$ is called the counting series of the class $\mathscr{A}$.

Now, let $\mathscr{A}$ and $\mathscr{B}$ be two combinatorial classes and let $a(x)$ and $b(x)$ be their respective counting series. Further, define the new class

$$\mathscr{C} := \mathscr{A} \times \mathscr{B} := \{(A, B) : A \in \mathscr{A}, B \in \mathscr{B}\},$$

where the size of a pair (A, B) is taken to be the sum of the sizes of A and B. Then the counting series $c(x)$ of $\mathscr{C}$ is easily seen to be

$$c(x) = a(x)b(x) = \sum_{n=0}^{\infty} \left(\sum_{k=0}^{n} a_k b_{n-k} \right) x^n.$$

2.2 Differentiation and Division

We have seen what formal power series are, and how addition and multiplication is defined for them. We will now turn to more sophisticated operations. First of all, although we have emphasized (and will continue to do so) that formal power series are to be distinguished from power series that represent analytic functions, we will need to (formally) *differentiate*.

In general, if R is a commutative ring and $D: R \to R$ is such that

$$D(a+b) = D(a) + D(b), \qquad D(a\,b) = D(a)b + a\,D(b)$$

for all $a, b \in R$, then D is called a (formal) *derivation* on R and the pair (R, D) is called a *differential ring*. A derivation on $\mathbb{K}[[x]]$ is naturally given by

$$D_x: \mathbb{K}[[x]] \to \mathbb{K}[[x]], \qquad D_x \sum_{n=0}^{\infty} a_n x^n := \sum_{n=0}^{\infty} a_{n+1}(n+1)x^n.$$

For example, we have

$$D_x \sum_{n=0}^{\infty} \frac{1}{n!} x^n = \sum_{n=0}^{\infty} \frac{n+1}{(n+1)!} x^n = \sum_{n=0}^{\infty} \frac{1}{n!} x^n.$$

We can also define a formal integration via

$$\int_x : \mathbb{K}[[x]] \to \mathbb{K}[[x]], \qquad \int_x \sum_{n=0}^{\infty} a_n x^n := \sum_{n=1}^{\infty} \frac{a_{n-1}}{n} x^n.$$

Note that we do not define integrals with arbitrary endpoints, but only integrals "from 0 to x". Also observe that so far we have made no assumptions on the ground field $\mathbb{K}$. In order to give a precise meaning to $a_{n+1}(n+1)$ and a_{n-1}/n, we have to say how natural numbers $n \in \mathbb{N}$ should be mapped into $\mathbb{K}$. This is easy: just map $n \in \mathbb{N}$ to $1 + 1 + \cdots + 1 \in \mathbb{K}$, where 1 is the multiplicative unit in $\mathbb{K}$ and the sum has exactly n terms. This is fine for the derivative, but for the integral it is troublesome when $\mathbb{K}$ is such that $1 + 1 + \cdots + 1$ can become zero, as for example in the case of finite fields. Let us be pragmatic and simply exclude these fields once and for all from consideration: throughout the rest of this book, $\mathbb{K}$ refers to a field where $1 + 1 + \cdots + 1 \neq 0$, regardless of the number of terms in the sum. Or equivalently: $\mathbb{K}$ is assumed to be a field of characteristic zero. Or equivalently: $\mathbb{K}$ contains a subfield which is isomorphic to $\mathbb{Q}$.

Although stated for arbitrary fields $\mathbb{K}$ (of characteristic zero), the definitions for derivative and integral given above are of course motivated from the respective theorems about converging power series in analysis, and as a consequence, they inherit some other properties known from analysis:

Theorem 2.3 *For all $a(x) \in \mathbb{K}[[x]]$ we have*

1. $D_x \int_x a(x) = a(x)$ *(Fundamental Theorem of Calculus I),*

2. $\int_x D_x a(x) = a(x) - a(0)$ *(Fundamental Theorem of Calculus II),*

3. $[x^n] a(x) = \dfrac{1}{n!} \left(D_x^n a(x) \right) \Big|_{x=0}$ *(Taylor's formula).*

The proof is left as Problem 2.5.

Viewed as operations on sequences, the derivation (Hadamard-)multiplies a sequence with $(n)_{n=0}^{\infty}$ and then *shifts* it by one index to the left. Integration acts on the coefficient sequence by first shifting it by one index to the right and then (Hadamard-)multiplying it by $(0, 1, \frac{1}{2}, \frac{1}{3}, \dots)$. A shift alone amounts to a multiplication by x:

$$x \sum_{n=0}^{\infty} a_n x^n = \sum_{n=1}^{\infty} a_{n-1} x^n \quad \longleftrightarrow \quad (0,1,0,0,\dots)(a_0, a_1, a_2, \dots) = (0, a_0, a_1, a_2, \dots).$$

A multiplication by n alone, without shift, can be obtained by combining derivative and forward shift:

$$xD_x \sum_{n=0}^{\infty} a_n x^n = \sum_{n=0}^{\infty} n a_n x^n$$

$$\longleftrightarrow \quad (0,1,2,3,\dots) \odot (a_0, a_1, a_2, \dots) = (0, a_1, 2a_2, 3a_3, \dots).$$

The operator xD_x is also a derivation on $\mathbb{K}[[x]]$.

Shifting backward drops out the constant term. Removing this coefficient first, all remaining terms in the series are divisible by x, and so a backward shift can be obtained by

$$\frac{1}{x}\left(\sum_{n=0}^{\infty} a_n x^n - a(0) \right) = \sum_{n=0}^{\infty} a_{n+1} x^n.$$

Note that without the correction term $-a(0)$, division by x is not defined, as there cannot be a term x^{-1} in a formal power series.

This leads to the question of *division:* under which circumstances does a formal power series $a(x)$ admit a multiplicative inverse in $\mathbb{K}[[x]]$? Obviously, the simple formal power series x does not have a multiplicative inverse, because $xb(x)$ has 0 as constant term and is therefore surely not equal to 1. On the other hand, the series

$$a(x) = \sum_{n=0}^{\infty} x^n$$

does have a multiplicative inverse, because

$$(1-x)a(x) = a(x) - xa(x) = \sum_{n=0}^{\infty} x^n - \sum_{n=1}^{\infty} x^n = 1.$$

It turns out that the constant term decides whether a multiplicative inverse of a series exists or not.

Theorem 2.4 (Multiplicative Inverse) *Let $a(x) \in \mathbb{K}[[x]]$. Then there exists a series $b(x) \in \mathbb{K}[[x]]$ with $a(x)b(x) = 1$ if and only if $a(0) \neq 0$.*

Proof. If $a(0) = 0$ then we can write $a(x) = x\tilde{a}(x)$ for some $\tilde{a}(x) \in \mathbb{K}[[x]]$. Then the constant term of $a(x)b(x) = x\tilde{a}(x)b(x)$ is zero for all $b(x) \in \mathbb{K}[[x]]$, and so in particular there cannot be a $b(x) \in \mathbb{K}[[x]]$ with $a(x)b(x) = 1$.

For the converse, let $a(x) = \sum_{n=0}^{\infty} a_n x^n$ and assume that $a(0) = a_0 \neq 0$. We construct inductively the coefficients of a series $b(x) = \sum_{n=0}^{\infty} b_n x^n$ with $a(x)b(x) = 1$. Start by setting $b_0 = 1/a_0$. Then we have $[x^0]a(x)b(x) = 1$ as desired. In order to satisfy

$$[x^1]a(x)b(x) = a_0 b_1 + a_1 b_0 \overset{!}{=} 0 = [x^1]1,$$

we set $b_1 = -a_1 b_0 / a_0$. In general, if the coefficients $b_0, \ldots, b_{n-1}$ of $b(x)$ are known, then the condition

$$[x^n]a(x)b(x) = \sum_{k=0}^{n} a_k b_{n-k} \overset{!}{=} 0 = [x^n]1$$

uniquely determines b_n as

$$b_n = -\frac{1}{a_0} \sum_{k=1}^{n} a_k b_{n-k}.$$

As the construction can be continued indefinitely, the proof is complete.   □

We use the usual notation $1/a(x)$ or $a(x)^{-1}$ to refer to the multiplicative inverse of $a(x)$.

The proof of the theorem tells us how to compute the first terms of $1/a(x)$ given the first terms of $a(x)$. But it does not tell us much about any properties the coefficients in a multiplicative inverse may have. For example, it is not at all evident (but true: Problem 2.18) that all the coefficients in the expansion

$$\left(\sum_{n=0}^{\infty} n! x^n \right)^{-1} = 1 - x - x^2 - 3x^3 - 13x^4 - 71x^5 - 461x^6 - 3447x^7 + \cdots$$

except for the constant term are negative. Quite different is the behavior of the coefficients c_n in the expansion

$$\frac{1}{1 - \log(1+x)} = 1 + x + \tfrac{1}{2}x^2 + \tfrac{1}{3}x^3 + \tfrac{1}{6}x^4 + \tfrac{7}{60}x^5 + \tfrac{19}{360}x^6 + \tfrac{3}{70}x^7 + \tfrac{5}{336}x^8 + \cdots,$$

with $\log(1+x)$ being the formal power series version of the logarithm defined as

$$\log(1+x) := \sum_{n=1}^{\infty} \frac{(-1)^{n+1}}{n} x^n = x - \tfrac{1}{2}x^2 + \tfrac{1}{3}x^3 - \tfrac{1}{4}x^4 + \tfrac{1}{5}x^5 - \tfrac{1}{6}x^6 + \cdots.$$

Although the first terms suggest that all the c_n are positive, this is not the case. It can be shown by a nontrivial argument that the sign alternates when n is large. The precise asymptotics is given by

$$c_n \sim (-1)^n \frac{1}{n \log(n)^2} \qquad (n \to \infty).$$

This example appears in a remarkable book of Polya [46].

The fact that a simple power series may have a complicated multiplicative inverse can sometimes be turned into an advantage. Some sequences which are otherwise hard to describe can be most easily expressed as the coefficient sequence of the multiplicative inverse of some simple power series. An example is the sequence of *Bernoulli numbers,* which we define as

$$B_n := n! [x^n] \frac{x}{\exp(x) - 1} \qquad (n \geq 0).$$

We have

$$\frac{x}{\exp(x) - 1} = 1 - \tfrac{1}{2}x + \tfrac{1}{12}x^2 - \tfrac{1}{720}x^4 + \tfrac{1}{30240}x^6 - \tfrac{1}{1209600}x^8 + \tfrac{1}{47900160}x^{10}$$

$$- \tfrac{691}{1307674368000}x^{12} + \tfrac{1}{74724249600}x^{14} - \tfrac{3617}{10670622842880000}x^{16} + \cdots,$$

with $\exp(x)$ being the formal power series version of the exponential function defined as

$$\sum_{n=0}^{\infty} \frac{1}{n!}x^n = 1 + x + \tfrac{1}{2}x^2 + \tfrac{1}{6}x^3 + \tfrac{1}{24}x^4 + \tfrac{1}{120}x^5 + \cdots.$$

The Bernoulli numbers appear in a number of different contexts, but they are not at all simple. We will see in Chap. 7 that they do not even belong to the wide class of holonomic sequences.

Clearly, if $b(x) \in \mathbb{K}[[x]]$ is such that $b(0) \neq 0$ then we can define the quotient $a(x)/b(x)$ as the product of $a(x)$ with the multiplicative inverse of $b(x)$. But in order for the quotient of two power series to be meaningful from a slightly more general point of view, it is not necessary that the power series in the denominator be invertible. We have already seen that it makes sense to speak about $(a(x) - a(0))/x$ because the power series in the numerator only involves terms $x, x^2, x^3, \ldots$ which all are divisible by x. This generalizes as follows. Define the *order* $\operatorname{ord} a(x)$ of a nonzero power series $a(x) = \sum_{n=0}^{\infty} a_n x^n \in \mathbb{K}[[x]]$ to be the smallest index n with $a_n \neq 0$. It is clear that $a(x)/x^{\operatorname{ord} a(x)}$ has a nonzero constant term and that $a(x)/x^k$ is a formal power series if and only if $k \leq \operatorname{ord} a(x)$. It is also easily seen that if $a(x), b(x) \in \mathbb{K}[[x]]$, then the quotient $a(x)/b(x)$ exists if and only if $\operatorname{ord} b(x) \leq \operatorname{ord} a(x)$. In this case, we have $\operatorname{ord} a(x)/b(x) = \operatorname{ord} a(x) - \operatorname{ord} b(x)$ and we say that $b(x)$ divides $a(x)$ or that $a(x)$ is a multiple of $b(x)$.

In order to have a multiplicative inverse for any nonzero power series, all we need is a multiplicative inverse x^{-1} of x. Then the multiplicative inverse of $x^r a(x)$ where $a(0) \neq 0$ is simply $x^{-r} a(x)^{-1}$, where $x^{-r} := (x^{-1})^r$ and $a(x)^{-1}$ is the multiplicative inverse of $a(x)$ in $\mathbb{K}[[x]]$. Series of the form $x^r a(x)$ where $r \in \mathbb{Z}$ and $a(x) \in \mathbb{K}[[x]]$ are called *(formal) Laurent series.* They correspond to sequences with index set $\{r, r+1, r+2, \ldots\} \subseteq \mathbb{Z}$. The set of all formal Laurent series over $\mathbb{K}$ is denoted by $\mathbb{K}((x))$. It is the quotient field of $\mathbb{K}[[x]]$.

2.3 Sequences of Power Series

Generally speaking, every operation that can be done with power series can also be done with formal power series, *as long as* that operation does not require adding up infinitely many numbers. For example, as seen in the previous section, the n-th coefficient of a multiplicative inverse is a linear combination of the coefficients from 0 to $n-1$ and thus only a finite sum. (There is no bound on the number of summands, though, as n grows indefinitely.) In contrast, it is in general not meaningful to "evaluate" a power series at some "point" $x \neq 0$, because this would lead to an infinite sum.

On the other hand, it is meaningful to replace x by $x^2 + x$ in any formal power series $a(x)$. To see why, observe first that

$$(x^2 + x)^n = x^n (x+1)^n = x^n \sum_{k=0}^{n} \binom{n}{k} x^k,$$

by the binomial theorem. If $a(x) = \sum_{n=0}^{\infty} a_n x^n$, then replacing x by $x^2 + x$ gives

$$\sum_{n=0}^{\infty} a_n x^n \sum_{k=0}^{n} \binom{n}{k} x^k = \sum_{n=0}^{\infty} \left(\sum_{k=0}^{n} \binom{n-k}{k} a_{n-k} \right) x^n.$$

As every coefficient in the new series is just a finite sum, we have a perfectly well-defined formal power series.

In order to generalize this reasoning, it is convenient to introduce the notion of a limit for sequences in $\mathbb{K}[[x]]$. Informally, we will say that two power series $a(x)$ and $b(x)$ are "close" if their first terms agree up to a high order, i.e., if $\mathrm{ord}(a(x) - b(x))$ is large. A sequence $(a_k(x))_{k=0}^{\infty}$ of formal power series is then said to converge to another formal power series $a(x)$ if the $a_k(x)$ get arbitrarily close to $a(x)$ in this sense. Formally, $(a_k(x))_{k=0}^{\infty}$ converges to $a(x)$ if and only if $\lim_{k \to \infty} \mathrm{ord}(a(x) - a_k(x)) = \infty$, i.e., if and only if

$$\forall n \in \mathbb{N} \ \exists k_0 \in \mathbb{N} \ \forall k \geq k_0 : \mathrm{ord}(a(x) - a_k(x)) > n.$$

In that case we write $a(x) = \lim_{k \to \infty} a_k(x)$, a notation which is justified because the limit of a convergent sequence of power series is unique.

If $a_k(x) = \sum_{n=0}^{\infty} a_{n,k} x^n$ and $a(x) = \sum_{n=0}^{\infty} a_n x^n$, then convergence of the sequence $(a_k(x))_{k=0}^{\infty}$ to $a(x)$ means that every coefficient sequence

$$a_{n,0}, \ a_{n,1}, \ a_{n,2}, \ a_{n,3}, \ \ldots$$

differs from a_n only for finitely many indices.

Theorem 2.5 Let $(a_n(x))_{n=0}^{\infty}$ and $(b_n(x))_{n=0}^{\infty}$ be two convergent sequences in $\mathbb{K}[[x]]$ and let $a(x) = \lim_{n \to \infty} a_n(x)$ and $b(x) = \lim_{n \to \infty} b_n(x)$. Then:

1. $(a_n(x) + b_n(x))_{n=0}^{\infty}$ is convergent and $\lim_{n \to \infty}(a_n(x) + b_n(x)) = a(x) + b(x)$.

2. $(a_n(x)b_n(x))_{n=0}^\infty$ *is convergent and* $\lim_{n\to\infty}(a_n(x)b_n(x)) = a(x)b(x)$.

The proof amounts to utilizing simple facts like

$$\text{ord}(a(x)+b(x)) \geq \min(\text{ord}\,a(x),\text{ord}\,b(x))$$
$$\text{and}\quad \text{ord}\,a(x)b(x) = \text{ord}\,a(x) + \text{ord}\,b(x)$$

for any two power series $a(x)$ and $b(x)$. We omit the details.

Using this notion of a limit, we can now proceed to define the composition of two formal power series. Let $a(x) = \sum_{n=0}^\infty a_n x^n, b(x) \in \mathbb{K}[[x]]$ be such that $b(0) = 0$. For every fixed $k \in \mathbb{N}$, $b(x)^k$ is a power series of order k or greater. Consider the sequence $(c_k(x))_{k=0}^\infty$ defined by

$$c_0(x) := a_0,$$
$$c_1(x) := a_0 + a_1 b(x),$$
$$c_2(x) := a_0 + a_1 b(x) + a_2 b(x)^2,$$

$$\vdots$$

$$c_k(x) := a_0 + a_1 b(x) + a_2 b(x)^2 + \cdots + a_k b(x)^k,$$

$$\vdots$$

Then the coefficients of $c_k(x)$ up to order k agree with the respective coefficients of all its successors, because all the $a_i b(x)^i$ added later have order greater than k. Therefore, the sequence $(c_k(x))_{k=0}^\infty$ converges to a formal power series $c(x) \in \mathbb{K}[[x]]$. We define the composition of $a(x)$ and $b(x)$ with $b(0) = 0$ as

$$a(b(x)) := \sum_{n=0}^\infty a_n b(x)^n := \lim_{k\to\infty} c_k(x) = c(x).$$

For example, for

$$a(x) = \exp(x) = \sum_{n=0}^\infty \frac{1}{n!}x^n = 1 + x + \tfrac{1}{2}x^2 + \tfrac{1}{6}x^3 + \tfrac{1}{24}x^4 + \tfrac{1}{120}x^5 + \tfrac{1}{720}x^6 + \cdots$$

$$b(x) = \exp(x) - 1 = \sum_{n=1}^\infty \frac{1}{n!}x^n = x + \tfrac{1}{2}x^2 + \tfrac{1}{6}x^3 + \tfrac{1}{24}x^4 + \tfrac{1}{120}x^5 + \tfrac{1}{720}x^6 + \cdots$$

we find

$$c_0(x) = \underline{1} + 0x + 0x^2 + 0x^3 + 0x^4 + 0x^5 + 0x^6 + 0x^7 + \cdots$$
$$c_1(x) = \underline{1 + x} + \tfrac{1}{2}x^2 + \tfrac{1}{6}x^3 + \tfrac{1}{24}x^4 + \tfrac{1}{120}x^5 + \tfrac{1}{720}x^6 + \tfrac{1}{5040}x^7 + \cdots$$
$$c_2(x) = \underline{1 + x + x^2} + \tfrac{2}{3}x^3 + \tfrac{1}{3}x^4 + \tfrac{2}{15}x^5 + \tfrac{2}{45}x^6 + \tfrac{4}{315}x^7 + \cdots$$
$$c_3(x) = \underline{1 + x + x^2 + \tfrac{5}{6}x^3} + \tfrac{7}{12}x^4 + \tfrac{41}{120}x^5 + \tfrac{61}{360}x^6 + \tfrac{73}{1008}x^7 + \cdots$$
$$c_4(x) = \underline{1 + x + x^2 + \tfrac{5}{6}x^3 + \tfrac{5}{8}x^4} + \tfrac{17}{40}x^5 + \tfrac{187}{720}x^6 + \tfrac{143}{1008}x^7 + \cdots$$

$$\vdots$$

The underlined parts where the coefficients are stable will grow larger as the index increases, and so we can record

$$\exp(\exp(x) - 1) = 1 + x + x^2 + \tfrac{5}{6}x^3 + \tfrac{5}{8}x^4 + \tfrac{13}{30}x^5 + \tfrac{203}{720}x^6 + \tfrac{877}{5040}x^7 + \cdots$$

for the limit. The numbers $\mathbf{B}_n := n![x^n]\exp(\exp(x) - 1)$ are called *Bell numbers*. They count the number of ways the set $\{1,\dots,n\}$ can be partitioned into non-empty disjoint sets. For example, $\mathbf{B}_4 = 4!\tfrac{5}{8} = 15$, because there are 15 partitions of $\{1,2,3,4\}$:

$\{1,2,3,4\}$,	$\{1\},\{2,3,4\}$,	$\{2\},\{1,3,4\}$,
$\{3\},\{1,2,4\}$,	$\{4\},\{1,2,3\}$,	$\{1,2\},\{3,4\}$,
$\{1,3\},\{2,4\}$,	$\{1,4\},\{2,3\}$,	$\{1\},\{2\},\{3,4\}$,
$\{1\},\{3\},\{2,4\}$,	$\{1\},\{4\},\{2,3\}$,	$\{2\},\{3\},\{1,4\}$,
$\{2\},\{4\},\{1,3\}$,	$\{3\},\{4\},\{1,2\}$,	$\{1\},\{2\},\{3\},\{4\}$.

The Bell numbers have no reasonable closed form.

We show next that the composition of power series is compatible with addition and multiplication.

Theorem 2.6 *For every fixed $u(x) \in \mathbb{K}[[x]]$ with $u(0) = 0$ the map*

$$\Phi_u \colon \mathbb{K}[[x]] \to \mathbb{K}[[x]], \qquad a(x) \mapsto a(u(x))$$

is a ring homomorphism.

Proof. Let $a(x) = \sum_{n=0}^{\infty} a_n x^n$ and $b(x) = \sum_{n=0}^{\infty} b_n x^n$. We show $\Phi_u(a(x))\Phi_u(b(x)) - \Phi_u(a(x)b(x)) = 0$; the proof of the corresponding statement for addition is straightforward. We have:

$$\Phi_u(a(x)b(x)) - \Phi_u(a(x))\Phi_u(b(x))$$

$$= \sum_{n=0}^{\infty}\left(\sum_{k=0}^{n} a_k b_{n-k}\right)u(x)^n - \sum_{n=0}^{\infty} a_n u(x)^n \sum_{n=0}^{\infty} b_n u(x)^n$$

$$= \lim_{N\to\infty}\sum_{n=0}^{N}\left(\sum_{k=0}^{n} a_k b_{n-k}\right)u(x)^n - \lim_{N\to\infty}\sum_{n=0}^{N} a_n u(x)^n \lim_{N\to\infty}\sum_{n=0}^{N} b_n u(x)^n$$

$$= \lim_{N\to\infty}\left(\sum_{n=0}^{N}\left(\sum_{k=0}^{n} a_k b_{n-k}\right)u(x)^n - \sum_{n=0}^{N} a_n u(x)^n \sum_{n=0}^{N} b_n u(x)^n\right) \qquad \text{(by Theorem 2.5)}$$

$$= -\lim_{N\to\infty}\sum_{n=N+1}^{2N}\left(\sum_{k=0}^{n} a_k b_{n-k}\right)u(x)^n.$$

This last limit is 0 (as we wish to show), because

$$\mathrm{ord}\left(\sum_{n=N+1}^{2N}\left(\sum_{k=0}^{n} a_k b_{n-k}\right)u(x)^n - 0\right) \geq \mathrm{ord}\,u(x)^{N+1} \geq N+1 \to \infty \quad (N \to \infty). \qquad \square$$

The key feature of Theorem 2.6 is that it allows us to keep our notations simple. In the first place, an expression like

$$\frac{1}{1-x^2}$$

is ambiguous: it may refer to the multiplicative inverse of $1-x^2$, or as well to the composition of $1/(1-x)$ with x^2. As the theorem implies that both interpretations lead to the same power series, there is no need for introducing a notational distinction among them.

Convergence arguments can also be used for justifying computations with infinite products of formal power series. For example, let

$$a_k(x) = \frac{1}{1-x^k} = 1 + x^k + x^{2k} + x^{3k} + x^{4k} + x^{5k} + \cdots \qquad (k \geq 1)$$

and consider the sequence $(c_k(x))_{k=0}^{\infty}$ where $c_k(x) = a_1(x)a_2(x)\cdots a_k(x)$. Since each $a_k(x)$ has the form $1 + x^k b_k(x)$ for some $b_k(x) \in \mathbb{K}[[x]]$, multiplication by $a_k(x)$ keeps the terms of small order as they are and only affects those of order k or more. That's why the sequence $(c_k(x))_{k=0}^{\infty}$ converges. Indeed, we have

$$c_1(x) = \underline{1} + x + x^2 + x^3 + x^4 + x^5 + x^6 + x^7 + \cdots$$
$$c_2(x) = \underline{1+x} + 2x^2 + 2x^3 + 3x^4 + 3x^5 + 4x^6 + 4x^7 + \cdots$$
$$c_3(x) = \underline{1+x+x^2} + 2x^3 + 3x^4 + 4x^5 + 5x^6 + 7x^7 + \cdots$$
$$\vdots$$

Eventually, we find the limit

$$c(x) = 1 + x + 2x^2 + 3x^3 + 5x^4 + 7x^5 + 11x^6 + 15x^7 + 22x^8 + 30x^9 + 42x^{10} + \cdots$$

The coefficients of this series also have a combinatorial meaning. The *partition number* $p_n := [x^n]c(x)$ is the number of ways in which the number n can be written as the sum of positive integers. For example $p_7 = 15$ because

$$7 = 6+1 = 5+2 = 5+1+1 = 4+3 = 4+2+1 = 4+1+1+1$$
$$= 3+3+1 = 3+2+2 = 3+2+1+1 = 3+1+1+1+1 = 2+2+2+1$$
$$= 2+2+1+1+1 = 2+1+1+1+1+1 = 1+1+1+1+1+1+1.$$

The examples indicate that the notion of convergence in $\mathbb{K}[[x]]$ can be used also to give a precise meaning to infinite sums and products. In analogy with the corresponding definitions, we say that an infinite sum

$$\sum_{n=0}^{\infty} a_n(x)$$

of power series *converges* if the sequence $(c_n(x))_{n=0}^{\infty}$ of partial sums $c_n(x) = \sum_{k=0}^{n} a_k(x)$ converges to some $c(x) \in \mathbb{K}[[x]]$ in the sense defined before. The convergence of an infinite product of power series is defined analogously.

Theorem 2.7 *Let $(a_n(x))_{n=0}^\infty$ be a sequence in $\mathbb{K}[[x]]$. Then the following statements are equivalent:*

1. $\displaystyle\lim_{n\to\infty} \operatorname{ord} a_n(x) = \infty$, 2. $\displaystyle\sum_{n=0}^\infty a_n(x)$ *converges*, 3. $\displaystyle\prod_{n=0}^\infty (1 + a_n(x))$ *converges*.

Proof. We show (1) $\Longleftrightarrow$ (2); the argument for (1) $\Longleftrightarrow$ (3) is Problem 2.11. Suppose $\lim_{n\to\infty} \operatorname{ord} a_n(x) = \infty$. Then

$$\forall\, n\; \exists\, k_0\; \forall\, k \ge k_0 : \operatorname{ord} a_k(x) > n.$$

Therefore, for $c_n(x) := \sum_{k=0}^n a_k(x)$ we have

$$\forall\, n\; \exists\, k_0\; \forall\, k \ge k_0 : \operatorname{ord}(c_{k+1}(x) - c_k(x)) > n.$$

For such n and k we therefore have $[x^n](c_k(x) - c_{k+1}(x)) = 0$, i.e., $[x^n]c_{k+1}(x) = [x^n]c_k(x)$, and the coefficient will remain fixed for all greater indices as well. For every n and some $k \ge k_0$, set $c_n := [x^n]c_k(x)$ and consider

$$c(x) := \sum_{n=0}^\infty c_n x^n.$$

Then by construction we have

$$\forall\, n\; \exists\, k_0\; \forall\, k \ge k_0 : \operatorname{ord}(c_k(x) - c(x)) > n,$$

and therefore the sequence $(c_n(x))_{n=0}^\infty$ converges, as was to be shown.

Conversely, suppose the infinite sum converges, i.e., suppose that the sequence $(c_n(x))_{n=0}^\infty$ with $c_n(x) := \sum_{k=0}^n a_k(x)$ converges to some formal power series $c(x) \in \mathbb{K}[[x]]$. Then

$$\forall\, n\; \exists\, k_0\; \forall\, k \ge k_0 : \operatorname{ord}(c_k(x) - c(x)) > n.$$

For such n and k, we have consequently that $\operatorname{ord}(c_{k+1}(x) - c(x)) > n$, and therefore $[x^n]c_k(x) = [x^n]c(x) = [x^n]c_{k+1}(x)$, and therefore

$$[x^n]a_{k+1}(x) = [x^n]\big(c_{k+1}(x) - c_k(x)\big) = [x^n]c_{k+1}(x) - [x^n]c_k(x) = 0,$$

and therefore $\operatorname{ord} a_{k+1}(x) > n$. Altogether, we have shown that

$$\forall\, n\; \exists\, k_0\; \forall\, k \ge k_0 : \operatorname{ord} a_k(x) > n,$$

as claimed. $\square$

2.4 The Transfer Principle

If $\mathbb{K} = \mathbb{R}$ or $\mathbb{K} = \mathbb{C}$ it is natural to ask for the relation between formal power series and analytic functions defined by power series. It is clear that not every formal power series corresponds to an analytic function. For example, the series

$$\sum_{n=0}^{\infty} n! x^n$$

considered earlier is a perfectly well-defined formal power series, but does not correspond to an analytic function as the ratio test gives

$$\lim_{n \to \infty} \frac{(n+1)! x^{n+1}}{n! x^n} = \lim_{n \to \infty} (n+1)x = \infty$$

for all $x \neq 0$. This example serves as a warning that a reasoning valid for formal power series may not have any meaning for analytic function.

If, on the other hand, two power series are identical as analytic functions, then they are identical also as formal power series:

Theorem 2.8 (Transfer Principle) *Let $a(z) = \sum_{n=0}^{\infty} a_n z^n$ and $b(z) = \sum_{n=0}^{\infty} b_n z^n$ be real or complex functions analytic in a non-empty open neighborhood U of zero. If $a(z) = b(z)$ for all $z \in U$, then $a_n = b_n$ for all $n \in \mathbb{N}$.*

Proof. Consider $c(z) = a(z) - b(z)$. Then $c(z)$ is analytic and identically zero on U, therefore, by Taylor's theorem about analytic functions,

$$c(z) = 0 + 0z + 0z^2 + 0z^3 + 0z^4 + \cdots.$$

Coefficient comparison gives

$$0 = [z^n]c(z) = [z^n]\big(a(z) - b(z)\big) = a_n - b_n \quad (n \in \mathbb{N}),$$

and the claim follows. $\square$

The transfer principle can be used for obtaining simple proofs of identities. For example, to see that

$$\exp(\log(1+x)) = 1+x$$

as formal power series, it suffices to note that the relation holds for the corresponding analytic functions. Doing the proof without transfer principle would require elaborate calculations.

The transfer principle is not only useful for proving identities. Reinterpretation of formal power series as analytic functions, where possible, it is also crucial for obtaining asymptotic estimates for the coefficients of a formal power series. Recall from analysis that a power series $\sum_{n=0}^{\infty} a_n z^n$ converges if $z \in \mathbb{C}$ is such that

$$|z| \limsup_{n \to \infty} |a_n|^{1/n} < 1,$$

and diverges for $z \in \mathbb{C}$ with

$$|z| \limsup_{n \to \infty} |a_n|^{1/n} > 1.$$

That is to say, the power series converges for all z in the open disk D around the origin with radius $r := 1/\limsup_{n \to \infty} |a_n|^{1/n}$ (understanding $1/\infty := 0$ and $1/0 := \infty$ here). It is usually (but not always!) possible to extend the analytic function

$$f \colon D \to \mathbb{C}, \qquad f(z) := \sum_{n=0}^{\infty} a_n z^n$$

beyond D, in the sense that for some connected open set $R \subseteq \mathbb{C}$ with $R \cap D$ open and $R \setminus D$ not empty, there exists an analytic function $F \colon R \to \mathbb{C}$ with $F(z) = f(z)$ for all $z \in D \cap R$. Such a function F is then called an *analytic continuation* of f. A point $\zeta \in \mathbb{C}$ which is not contained in any open set $R \subseteq \mathbb{C}$ to which f may be analytically continued in this sense is called a *singularity* of f. For example, the function $z \mapsto 1/(1+z)$ has a singularity at $z = -1$. It can be shown that the boundary of D contains at least one such point. Conversely, if a complex function is analytic in a neighborhood of the origin, then the radius of convergence of its Taylor series expansion at the origin is determined by the absolute value of the singularity that is closest to the origin.

The singularities of an analytic function that lie on the boundary of the disk of convergence are called *dominant singularities*. They govern the asymptotics of the coefficient sequence $(a_n)_{n=0}^{\infty}$ of the power series expansion $\sum_{n=0}^{\infty} a_n z^n$. Indeed, the $\limsup$ relation above implies directly that if there is a dominant singularity at a point z_0, then we will have $|a_n| < (\frac{1}{|z_0|} + \varepsilon)^n$ for all $\varepsilon > 0$ and all sufficiently large n, and that this estimate is sharp in the sense that we will also have $|a_n| > (\frac{1}{|z_0|} - \varepsilon)^n$ for infinitely many n. As an example, consider the coefficients of

$$\sum_{n=0}^{\infty} a_n x^n := \frac{x}{\sin(x)} = 1 + \tfrac{1}{6}x^2 + \tfrac{7}{360}x^4 + \tfrac{31}{15120}x^6 + \tfrac{127}{604800}x^8 + \tfrac{73}{3421440}x^{10} + \cdots,$$

where $\sin(x)$ is the formal power series version in $\mathbb{K}[[x]]$ of the real or complex sine function, defined as

$$\sin(x) := \sum_{n=0}^{\infty} \frac{(-1)^n}{(2n+1)!} x^{2n+1} = x - \tfrac{1}{6}x^3 + \tfrac{1}{120}x^5 - \tfrac{1}{5040}x^7 + \tfrac{1}{362880}x^9 + \cdots.$$

In order to obtain information about the asymptotic behavior of the a_n, we consider the function

$$f \colon \mathbb{C} \setminus \{\ldots, -2\pi, -\pi, 0, \pi, 2\pi, \ldots\} \to \mathbb{C}, \quad f(z) = \frac{z}{\sin(z)}.$$

This function has a removable singularity at $z = 0$, and the extension $f(0) := 1$ turns it into a function that is analytic in a neighborhood of the origin. The radius of convergence of its power series expansion around zero is $r = \pi$, and there are two dominant singularities at $\pm\pi$. Therefore, the coefficients a_n stay asymptotically below $(\frac{1}{\pi}+\varepsilon)^n$ for every choice of $\varepsilon>0$.

Finer information on the asymptotics can be obtained from the structure of the dominant singularities. In our example, the two dominant singularities are simple poles with residues $-\pi$ (at $z=\pi$) and π (at $z=-\pi$), respectively. Therefore, if we define the auxiliary function

$$h\colon \mathbb{C}\setminus\{-\pi,\pi\}\to\mathbb{C},\quad h(z):=\frac{\pi}{\pi-z}-\frac{\pi}{-\pi-z},$$

then $f(z)-h(z)$ has removable singularities at $\pm\pi$ and can thus be extended to an analytic function in a disk of radius 2π around the origin. This implies that the coefficients of the formal power series

$$\frac{x}{\sin(x)}-\left(\frac{\pi}{\pi-x}-\frac{\pi}{-\pi-x}\right)$$

are bounded in absolute value by $(\frac{1}{2\pi}+\varepsilon)^n$ when n is large. In other words,

$$\left|a_n-(\pi^{-n}+(-\pi)^{-n})\right|<\left(\frac{1}{2\pi}+\varepsilon\right)^n\qquad(n\to\infty)$$

for every $\varepsilon>0$. Since $a_n=0$ when n is odd, it follows that $a_{2n}\sim 2\pi^{-2n}$ as $n\to\infty$.

Such considerations can be driven much further. In fact, it is possible to obtain accurate asymptotic information for a sequence from the asymptotic behavior of its generating function near its dominant singularities. The book of Flajolet and Sedgewick [21] contains a comprehensive introduction to the relevant techniques.

2.5 Multivariate Power Series

Most of what was said so far about formal power series over a field $\mathbb{K}$ generalizes in a natural way to formal power series with coefficients in a commutative ring R. In particular, $R[[x]]$ is an integral domain if R is, and a power series $a(x)\in R[[x]]$ has a multiplicative inverse in $R[[x]]$ whenever its constant term $a(0)$ has a multiplicative inverse in R (Problem 2.6). We can therefore define the ring of formal power series in two variables x,y simply by setting $\mathbb{K}[[x,y]]:=\mathbb{K}[[x]][[y]]$ ($\cong \mathbb{K}[[y]][[x]]$), and analogously for more variables. Formal power series in one, two, or many variables are also called univariate, bivariate, or multivariate, respectively.

Not too surprisingly, formal power series in several variables are useful for working with sequences in several variables. For example, consider the simple power series

$$1-x-xy=(1-x+0x^2+0x^3+\cdots)+(0-x+0x^2+\cdots)y+0y^2+0y^3+\cdots.$$

This series, considered as a series in y over $\mathbb{K}[[x]]$, has the constant term $1-x$, which in turn, as a series in x over $\mathbb{K}$, has the constant term 1. Therefore, $1-x$ is invertible in $\mathbb{K}[[x]]$ and therefore $1 - x - xy$ is invertible in $\mathbb{K}[[x,y]]$. What is the inverse? We can determine its coefficients in the same way as for power series in one variable:

$$
\begin{aligned}
\frac{1}{1-x-xy} &= \frac{1}{1-x} + \frac{x}{(1-x)^2}y + \frac{x^2}{(1-x)^3}y^2 + \frac{x^3}{(1-x)^4}y^4 + \cdots \\
&= (1 + x + x^2 + x^3 + x^4 + x^5 + \cdots) \\
&\quad + (x + 2x^2 + 3x^3 + 4x^4 + 5x^5 + \cdots)y \\
&\quad + (x^2 + 3x^3 + 6x^4 + 10x^5 + \cdots)y^2 \\
&\quad + (x^3 + 4x^4 + 10x^5 + \cdots)y^3 \\
&\quad + (x^4 + 5x^5 + \cdots)y^4 \\
&\quad + \cdots.
\end{aligned}
$$

The bivariate sequence emerging here is the ubiquitous *binomial coefficient* sequence $\binom{n}{k}$ which we will study in greater detail in Sect. 5.1. For the moment, we record

$$
\frac{1}{1-x-xy} = \sum_{n,k=0}^{\infty} \binom{n}{k} x^n y^k
$$

as its bivariate generating function.

Like in the univariate case, operations on multivariate power series correspond to operations on the (multivariate) coefficient sequence. For example, multiplication by x corresponds to a shift in n and multiplication by y corresponds to shift in k. Our result about the bivariate generating function of the binomial coefficients is therefore just a reformulation of the Pascal triangle relation:

$$
(1 - x - xy) \sum_{n,k=0}^{\infty} \binom{n}{k} x^n y^k = 1 \implies \binom{n}{k} - \binom{n-1}{k} - \binom{n-1}{k-1} = 0 \quad (n,k > 0).
$$

It may at times be of interest to view a bivariate power series $a(x,y)$ as a power series in one variable with coefficients being power series in the other. For example, regarding $1/(1 - x - xy)$ as power series in y with coefficients in $\mathbb{K}[[x]]$, we find from the geometric series the representation

$$
\frac{1}{1-x-xy} = \frac{1/(1-x)}{1 - xy/(1-x)} = \frac{1}{1-x} \sum_{k=0}^{\infty} \left(\frac{xy}{(1-x)} \right)^k = \sum_{k=0}^{\infty} \frac{x^k}{(1-x)^{k+1}} y^k,
$$

so

$$
[y^k] \frac{1}{1-x-xy} = \frac{x^k}{(1-x)^{k+1}} \in \mathbb{K}[[x]] \quad \text{and} \quad \binom{n}{k} = [x^n] \frac{x^k}{(1-x)^{k+1}}.
$$

In the other direction, regarding $1/(1 - x - xy)$ as a power series in x with coefficients in $\mathbb{K}[[y]]$, we find

$$
\frac{1}{1-x-xy} = \frac{1}{1 - (1+y)x} = \sum_{n=0}^{\infty} (1+y)^n x^n,
$$

so

$$[x^n]\frac{1}{1-x-xy}=(1+y)^n\in\mathbb{K}[[y]]\quad\text{and}\quad\binom{n}{k}=[y^k](1+y)^n.$$

The take-home message from these calculations is that already a single additional variable gives a lot of additional possibilities to look at a series and its coefficients.

Let us return to general considerations and investigate under which circumstances the infinite sum or product of a sequence of multivariate power series is meaningful. We have seen that all we need to ensure is that no infinite number of field elements gets summed or multiplied up. If we define the *total degree* of a term $x_1^{n_1}x_2^{n_2}\cdots x_m^{n_m}$ as the sum $n_1+n_2+\cdots+n_m$, then every nonzero formal power series $a(x_1,\ldots,x_m)\in\mathbb{K}[[x_1,\ldots,x_m]]$ will have some term(s) of minimal total degree whose coefficient is nonzero. We define this minimal degree as the (total) order of $a(x_1,\ldots,x_m)$. With this definition, the results of Sect. 2.3 carry over to the multivariate setting. In particular, if $(a_n(x_1,\ldots,x_m))_{n=0}^\infty$ is a sequence of multivariate formal power series, then

$$\sum_{n=0}^{\infty}a_n(x_1,\ldots,x_m)\quad\text{and}\quad\prod_{n=0}^{\infty}a_n(x_1,\ldots,x_m)$$

exist if and only if the total order of $a_n(x_1,\ldots,x_m)$ is unbounded as n goes to infinity. As a consequence, we can form the composed power series

$$\exp(x+y)=\sum_{n=0}^{\infty}\frac{1}{n!}(x+y)^n\in\mathbb{K}[[x,y]],$$

because $x+y\in\mathbb{K}[[x,y]]$ has order one. Observe that the existence of $\exp(x+y)$ does not follow directly from the univariate theorem upon considering $x+y$ as element of $\mathbb{K}[[x]][[y]]$, because in that sense, $x+y$ has the nonzero constant term x.

Of course, differentiation can be defined for multivariate power series just as for univariate ones. In the multivariate setting, we have a partial derivative for each variable that acts on a power series in the expected way. For example, we have

$$D_y\sum_{n,m=0}^{\infty}a_{n,m}x^ny^m:=\sum_{n,m=0}^{\infty}a_{n,m+1}(m+1)x^ny^m.$$

After addition, multiplication, differentiation, division, composition, and infinite sums and products, we now turn to one final way of defining a power series in terms of others: a formal version of the implicit function theorem that allows us to define power series as solutions of power series equations. For simplicity, we formulate it for the case of defining a univariate power series by a bivariate equation.

Theorem 2.9 (Implicit Function Theorem) *Let $a(x,y)\in\mathbb{K}[[x,y]]$ be such that*

$$a(0,0)=0\quad\text{and}\quad(D_ya)(0,0)\neq 0.$$

Then there exists a unique formal power series $f(x)\in\mathbb{K}[[x]]$ with $f(0)=0$ such that $a(x,f(x))=0$.

Proof. Write $a(x,y) = \sum_{n=0}^{\infty} a_n(x)y^n$ with $a_n(x) \in \mathbb{K}[[x]]$. The conditions on $a(x,y)$ translate into $a_0(0) = 0$ and $a_1(0) \neq 0$. We will show how to construct inductively the coefficients of a solution $f(x) = \sum_{n=1}^{\infty} f_n x^n$. By $a(0,0) = 0$ and $f(0) = 0$, we get $[x^0]a(x, f(x)) = 0$ for free. Next,

$$[x^1] \sum_{n=0}^{\infty} a_n(x)f(x)^n \overset{!}{=} 0$$

forces

$$[x^1]\big(a_0(x) + a_1(x)f(x)\big) = [x^1]a_0(x) + f_1 a_1(0) \overset{!}{=} 0,$$

so $f_1 = -[x^1]a_0(x)/a_1(0)$, where the division is allowed by assumption.

For the general case, assume that coefficients $f_0, \ldots, f_{n-1}$ are uniquely determined by the equation. Then

$$[x^n] \sum_{n=0}^{\infty} a_n(x)f(x)^n \overset{!}{=} 0$$

forces

$$[x^n]\big(a_0(x) + a_1(x)f(x) + a_2(x)f(x)^2 + \cdots + a_n(x)f(x)^n\big) \overset{!}{=} 0.$$

Now observe that

- $[x^n]a_1(x)f(x) = a_1(0)f_n + $ terms depending on $f_0, \ldots, f_{n-1}$ only;
- $[x^n]a_k(x)f(x)^k$ for $k > 1$ depends on $f_0, \ldots, f_{n-k+1}$ only because $f_0 = 0$.

Therefore, the previous condition forces the unique solution

$$f_n = (\cdots \text{ terms depending only on } f_0, \ldots, f_{n-k+1} \cdots)/a_1(0).$$

As n was arbitrary, this process can be continued indefinitely and the proof is complete.   □

In connection with Theorem 2.6, the implicit function theorem implies that a substitution $x \mapsto u(x)$ is an isomorphism of $\mathbb{K}[[x]]$ if $\operatorname{ord} u(x) = 1$.

The implicit function theorem can also be useful for writing down simple definitions of otherwise complicated power series. For example, we can define a formal power series $W(x) \in \mathbb{Q}[[x]]$ with $W(0) = 0$ by the equation

$$W(x)\exp(W(x)) = x.$$

To this end, consider $a(x,y) = y\exp(y) - x \in \mathbb{Q}[[x,y]]$. Since $a(0,0) = 0$ and $D_y a(x,y)|_{x,y=0} = (1+y)\exp(y)|_{x,y=0} = 1 \neq 0$, such a power series $W(x)$ exists and is according to Theorem 2.9 uniquely determined by the equation. As in former situations, the proof provides us with a method for determining the first coefficients. We find:

$$W(x) = x - x^2 + \tfrac{3}{2}x^3 - \tfrac{8}{3}x^4 + \tfrac{125}{24}x^5 - \tfrac{54}{5}x^6 + \tfrac{16807}{720}x^7 + \cdots.$$

It is not hard to imagine that power series defined in this way may lead to pretty wild coefficient sequences. Indeed, typically they do. Incidentally, for the example we have chosen, the coefficients admit a simple closed form:

$$W(x) = \sum_{n=0}^{\infty} \frac{(-n)^{n-1}}{n!} x^n.$$

The series $W(x)$ is known as the *Lambert W-function*, and it is interesting in combinatorics mainly because the numbers n^{n-1} count the number of labelled rooted trees with n nodes.

2.6 Truncated Power Series

As power series are infinite objects, exact computations with them are only possible when attention is restricted to power series that can be described by some sort of finite expression. In the remaining chapters of this book we will see several ways to do this. For the unrestricted domain of all power series, computations are only possible if we resort to some kind of *approximation*. We have seen in the foregoing sections that we can obtain the first few terms of, say, the product of two power series given only their first few terms. This suggests one to consider the first few terms of a series as an approximation to the entire series. The more terms are included, the finer we consider the approximation. To be precise, if $a(x) = \sum_{n=0}^{\infty} a_n x^n$, then we call

$$a_k(x) = a_0 + a_1 x + a_2 x^2 + \cdots + a_k x^k$$

an *approximation of order k* to $a(x)$. Note that $a(x)$ and $a_k(x)$ are close in the sense previously defined when k is large: we have $\mathrm{ord}(a(x) - a_k(x)) > k$. We call $a_k(x)$ a *truncated power series*. It results from a truncation of $a(x)$ at order k. The more careful alternative notation

$$a_k(x) = a_0 + a_1 x + a_2 x^2 + \cdots + a_k x^k + \mathrm{O}(x^{k+1})$$

can be used for making the truncation order explicit. This is yet another example for a notation which is inspired from analysis but which is used here in a purely algebraic sense.

Approximations of order k can be computed for product, derivative, integral, quotient, and composition of power series given by approximations of order k. This is what we have been already doing in the examples above. Naive algorithms follow directly from the respective definitions or theorems, more efficient ones can be found in Sect. 4.7 of [34].

The first terms of a sequence play an important role in the study of combinatorial problems. It is often so that long before anything else about a problem is understood, we can already determine the beginning of a sequence. Starring long enough at the first terms of a sequence can be the first step towards the solution of the general problem. For example, consider the number of plane binary trees with n internal nodes:

1, 1, 2, 5, 14, 42, 132, 429, 1430, 4862, 16796, 58786, 208012, 742900,
    2674440, 9694845, 35357670, 129644790, 477638700, 1767263190,
    6564120420, 6564120420, 24466267020, 91482563640, ...

Figure 2.1 shows the 14 trees with four internal nodes.

Is there any pattern in these numbers? If no pattern is apparent to the naked eye, we can look up the sequence in a table such as Neil Sloane's Online Encyclopedia of Integer Sequences (http://www.research.att.com/~njas/sequences/). This is a database containing the first terms of more than 150000 sequences of integers, together with all sorts of information known about them. If we type the numbers above into the system, we will learn that they are known as the Catalan numbers C_n, that they appear in a vast number of different combinatorial contexts, and that there is a variety of formulas by which they may be described.

For sequences not known to the database, we can try to guess some properties of the sequence computationally. For example, we can try to estimate the asymptotic growth. It is often so, even when a power series is far from having a simple representation, and when there is no hope for a simple description of the coefficient sequence, that the asymptotic behavior of the coefficients can still be stated in simple terms. If we are willing to make the assumption that a sequence grows like $p(n)\phi^n$ for some constant ϕ and some polynomial $p(n)$, then we can obtain an estimate for ϕ

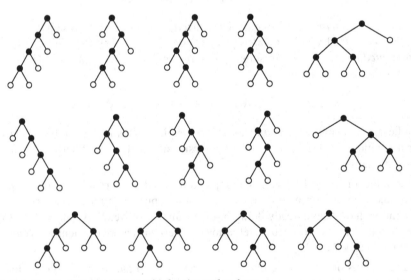

Fig. 2.1 The 14 plane binary trees with four internal nodes

from the first terms of the sequence. For example, suppose we know nothing about the sequence $(C_n)_{n=0}^\infty$ of Catalan numbers except the twenty first terms listed above. If C_n is asymptotically equal to a term $p(n)\phi^n$ with $p(n+1)/p(n) \to 1$ $(n \to \infty)$, then we should have that $q_n := C_{n+1}/C_n \to \phi$ for $n \to \infty$. For $n = 10,\ldots,19$, the quotient evaluates to

$$3.5, \ 3.53846, \ 3.57143, \ 3.6, \ 3.625, \ 3.64706, \ 3.66667, \ 3.68421, \ 3.7, \ 3.71429.$$

Could well be that this converges. But to what limit? There is a simple trick due to Richardson that turns a convergent sequence into a sequence converging more quickly to the same limit: Instead of q_n, consider $2q_{2n} - q_n$. See Problem 2.19 for an explanation. Indeed, for $n = 1,\ldots,9$, this evaluates to

$$3., \ 3.5, \ 3.7, \ 3.8, \ 3.85714, \ 3.89286, \ 3.91667, \ 3.93333, \ 3.94545,$$

and it is not hard to guess from these values that $\phi = 4$, which is indeed correct. (In fact, we have $C_n \sim 4^n/\sqrt{\pi n^3}$, as we shall see later.) When hundreds or thousands of terms are available, ϕ can usually be determined with an accuracy of a dozen decimal digits or so.

Finally, a truncated series is also useful for coming up with plausible guesses for equations the full series may possibly satisfy. Let us consider again the Catalan numbers. We want to find an equation that the power series

$$C(x) = 1 + x + 2x^2 + 5x^3 + 14x^4 + 42x^5 + 132x^6 + 429x^7 + 1430x^8 + 4862x^9 + \cdots$$

possibly obeys, assuming that we know the series only truncated at order ten. Searching for an equation, we have to decide on some type of equations. There are several possibilities. We may ask for instance whether there is an equation of the form

$$(u_0 + u_1 x + u_2 x^2)C(x) - (u_3 + u_4 x + u_5 x^2) = 0.$$

If so, then it must still hold to order k if we replace $C(x)$ by an approximation of $C(x)$ to order k. Plugging into the equation the first terms we know about $C(x)$, we obtain

$$(u_0 - u_3) + (2u_0 + u_1 - u_4)x + (5u_0 + 2u_1 + u_2 - u_5)x^2 + (14u_0 + 5u_1 + 2u_2)x^3$$

$$+ (42u_0 + 14u_1 + 5u_2)x^4 + (132u_0 + 42u_1 + 14u_2)x^5 + (429u_0 + 132u_1 + 42u_2)x^6$$

$$+ (1430u_0 + 429u_1 + 132u_2)x^7 + (4862u_0 + 1430u_1 + 429u_2)x^8$$

$$+ (16796u_0 + 4862u_1 + 1430u_2)x^9 + \cdots \overset{!}{=} 0.$$

Comparing coefficients to zero gives a linear system for the u_i which in our case has only the trivial solution $u_0 = \cdots = u_5 = 0$. Therefore, it can be concluded that $C(x)$ does not satisfy any equation of the attempted form.

What about a nonlinear equation? If we search in the same way for an equation of the form

$$(u_0 + u_1 x + u_2 x^2) + (u_3 + u_4 x + u_5 x^2)C(x) + (u_6 + u_7 x + u_8 x^2)C(x)^2 = 0,$$

we will find that equating all the coefficients to zero gives the nontrivial solution

$$(u_0, u_1, u_2, u_3, u_4, u_5, u_6, u_7, u_8) = (1, 0, 0, -1, 0, 0, 0, 1, 0)$$

that corresponds to the equation

$$1 - C(x) + xC(x)^2 = 0.$$

Does $C(x)$ really satisfy this equation? This equation cannot be answered from our reasoning alone (and in fact from *no* reasoning whatsoever if only some finitely many terms of $C(x)$ are known). The calculations we have done only assert that

$$1 - C(x) + xC(x)^2 = 0 + 0x + 0x^2 + 0x^3 + 0x^4 + 0x^5 + 0x^6 + 0x^7 + 0x^8 + 0x^9 + \cdots$$

and we cannot be sure without further reasoning that all the terms hidden in the dots will have zero as coefficient. But even so, an automatically "guessed" equation can be extremely valuable, as it does provide a strong hint towards what we may want to prove about the series at hand. Indeed, we will confirm in Chap. 6 that the equation we found is correct. Power series satisfying equations of this form will be called *algebraic*. If $C(x)$ had satisfied an equation of the form we tried first, we would have said that $C(x)$ is *rational*. Such power series are the topic of Chap. 4. We can also use an ansatz with undetermined coefficients for discovering potential linear *differential equations* with polynomial coefficients that a power series may satisfy. Power series satisfying such equations are called *holonomic* and studied in Chap. 7.

2.7 Problems

Problem 2.1 Prove Theorem 2.1.

Problem 2.2 Prove that $\mathbb{K}[[x]]$ is a Euclidean domain. What is the greatest common divisor of two power series?

Problem 2.3 Prove the exponential law for formal power series,

$$\exp(ax)\exp(bx) = \exp((a+b)x) \qquad (a, b \in \mathbb{K}).$$

Problem 2.4 Review the analysis of Quicksort presented in Sect. 1.4. Which operations performed there, if any, cannot be justified in the algebraic framework of formal power series?

Problem 2.5 Prove Theorem 2.3.

Problem 2.6 Let R be a commutative ring. Prove that $R[[x]]$ is an integral domain whenever R is, and that $a(x) \in R[[x]]$ has a multiplicative inverse in $R[[x]]$ if and only if $a(0)$ has a multiplicative inverse in R.

Problem 2.7 If (R,D) is a differential ring, then

$$C(R) := \{c \in R : D(c) = 0\}$$

is called the set of *constants* in (R,D). Prove that $C(R)$ is a subring of R.

Problem 2.8 Find a closed form for the coefficients in the multiplicative inverses of $(1 - 2x)^2$, $(1 - x)^3$ and $\exp(2x)$.

Problem 2.9 We say that a sequence $(a_n(x))_{n=0}^{\infty}$ in $\mathbb{K}[[x]]$ is a *Cauchy-sequence* if

$$\forall\, n \in \mathbb{N}\ \exists\, k_0 \in \mathbb{N}\ \forall\, k,l \geq k_0 : \operatorname{ord}\big(a_k(x) - a_l(x)\big) > n.$$

Prove that $(a_n(x))_{n=0}^{\infty}$ is a Cauchy-sequence if and only if it converges in the sense as defined in the text.

(*Hint:* Recycle the proof for the corresponding statement about sequences of real numbers.)

Problem 2.10 The infinite continued fraction

$$\cfrac{x}{1 - \cfrac{x}{1 - \cfrac{x}{1 - \cdots}}} \in \mathbb{K}[[x]]$$

is defined as the limit of the sequence

$$x,\ \cfrac{x}{1-x},\ \cfrac{x}{1 - \cfrac{x}{1-x}},\ \cfrac{x}{1 - \cfrac{x}{1 - \cfrac{x}{1-x}}},\ \cfrac{x}{1 - \cfrac{x}{1 - \cfrac{x}{1 - \cfrac{x}{1-x}}}},\ \ldots$$

Prove that this limit exists.

(*Hint:* You may use without proof that for any two power series $u(x), v(x) \in \mathbb{K}[[x]]$ with $u(0) \neq 0$ and $v(0) \neq 0$ and for every $k \in \mathbb{N}$ we have that $\operatorname{ord}(u(x) - v(x)) > k$ implies $\operatorname{ord}(\frac{1}{u(x)} - \frac{1}{v(x)}) > k$.)

Problem 2.11 Prove the equivalence (1) $\Longleftrightarrow$ (3) of Theorem 2.7.

Problem 2.12 Prove that the Bernoulli numbers B_n satisfy the equation

$$\sum_{k=0}^{n} \binom{n+1}{k} B_k = 0 \quad (n \geq 1).$$

Problem 2.13 Apply the reasoning of Sect. 2.4 to the formal power series

$$\sum_{n=0}^{\infty} \frac{B_n}{n!} x^n$$

in order to prove the asymptotic estimate

$$B_{2n} \sim 2(-1)^{n+1} \frac{(2n)!}{(2\pi)^{2n}} \quad (n \to \infty)$$

for the Bernoulli numbers.

Problem 2.14 Prove that the Bell numbers $\mathbf{B}_n$ satisfy the equation

$$\mathbf{B}_{n+1} = \sum_{k=0}^{n} \binom{n}{k} \mathbf{B}_k \quad (n \geq 1).$$

Problem 2.15 (Euler transform) Use $[x^n] \frac{x^k}{(1-x)^{k+1}} = \binom{n}{k}$ to show that for all $a(x) \in \mathbb{K}[[x]]$, we have

$$[x^n] \frac{1}{1-x} a\left(\frac{x}{x-1}\right) = \sum_{k=0}^{n} (-1)^k \binom{n}{k} [x^k] a(x) \quad (n \geq 0).$$

Problem 2.16 Let $a(x), b(x) \in \mathbb{K}[[x]]$ with $b(0) = 0$, write $a'(x) = D_x a(x)$ and $b'(x) = D_x b(x)$. Prove the chain rule $D_x a(b(x)) = a'(b(x)) b'(x)$.

(*Hint:* You may use without proof that derivation of formal power series commutes with limits: $D_x \lim_{k \to \infty} a_k(x) = \lim_{k \to \infty} D_x a_k(x)$.)

Problem 2.17 Using the method outlined in the proof of Theorem 2.9, determine the first ten coefficients of the formal power series $f(x) \in \mathbb{K}[[x]]$ with $f(0) = 1$ and

$$f(x)^2 = x^2 + 1$$

Problem 2.18 Check out what Sloane's database has to say about Bell numbers and about the coefficients of $\left(\sum_{n=0}^{\infty} n! x^n\right)^{-1}$.

Problem 2.19 Richardson's convergence acceleration rests on the assumption that a sequence $(q_n)_{n=0}^{\infty}$ converging to a limit c can be written in the form

$$q_n = c\left(1 + \frac{\alpha_1}{n} + \frac{\alpha_2}{n^2} + \frac{\alpha_3}{n^3} + \frac{\alpha_4}{n^4} + \cdots\right) \qquad (n \to \infty),$$

for in that case, the calculation

$$\ominus \left\{ \begin{array}{l} 2q_{2n} = 2c\left(1 + \dfrac{\alpha_1}{2n} + \dfrac{\alpha_2}{4n^2} + \dfrac{\alpha_3}{8n^3} + \dfrac{\alpha_4}{16n^3} + \dfrac{\alpha_5}{32n^4} + \cdots\right) \\[2ex] q_n = c\left(1 + \dfrac{\alpha_1}{n} + \dfrac{\alpha_2}{n^2} + \dfrac{\alpha_3}{n^3} + \dfrac{\alpha_4}{n^3} + \dfrac{\alpha_5}{n^4} + \cdots\right) \end{array} \right.$$

$$2q_{2n} - q_n = c\left(1 + 0 - \frac{3\alpha_2}{4n^2} - \frac{7\alpha_3}{8n^3} - \frac{15\alpha_4}{16n^3} - \frac{31\alpha_5}{32n^5} - \cdots\right)$$

suggests that the new sequence converges like $1/n^2$ instead of just like $1/n$, because the term α_1/n is eliminated.

Generalize this reasoning so as to eliminate both the terms α_1/n and α_2/n^2. What estimation for ϕ can be obtained from the first twenty Catalan numbers using the improved scheme?

Problem 2.20 Which equations do the power series below seem to satisfy?

1. $x + x^2 + 2x^3 + 3x^4 + 5x^5 + 8x^6 + 13x^7 + 21x^8 + 34x^9 + 55x^{10} + \cdots$

2. $1 - \frac{1}{2}x - \frac{5}{8}x^2 - \frac{5}{16}x^3 - \frac{45}{128}x^4 - \frac{95}{256}x^5 - \frac{465}{1024}x^6 - \frac{1165}{2048}x^7 - \frac{24445}{32768}x^8 - \frac{65595}{65536}x^9$
   $- \frac{359915}{262144}x^{10} - \frac{1003315}{524288}x^{11} - \frac{11342185}{4194304}x^{12} + \cdots$

3. $1 + x - \frac{1}{2}x^2 - \frac{11}{6}x^3 - \frac{47}{24}x^4 - \frac{139}{120}x^5 - \frac{89}{720}x^6 + \frac{3529}{5040}x^7 + \frac{47489}{40320}x^8 + \frac{486793}{362880}x^9$
   $+ \frac{4628431}{3628800}x^{10} + \frac{4628431}{3628800}x^{10} + \frac{42461981}{39916800}x^{11} + \cdots$

Chapter 3
Polynomials

In this chapter we start to study restricted classes of power series. The most simple classes are the most restricted ones, and so we will start with those. The power series considered in this chapter have in common that they all can be described by polynomials. Polynomials are not only interesting as a class in its own right, but they will also be employed as building blocks in the construction of the more elaborate classes that are considered in the following chapters.

3.1 Polynomials as Power Series

For $n \geq 0$, let a_n be the number of unlabeled graphs with four vertices and n edges. Then, according to Fig. 3.1, we have the generating function

$$a(x) = \sum_{n=0}^{\infty} a_n x^n = 1 + x + 2x^2 + 3x^3 + 2x^4 + x^5 + x^6.$$

There is no graph with four vertices and more than six edges, so we have $a_n = 0$ for all $n > 6$. This feature distinguishes the sequence $(a_n)_{n=0}^{\infty}$ from all the sequences we have seen in examples so far. We can regard $(a_n)_{n=0}^{\infty}$ as a *finite* sequence.

Power series of finite sequences have a lot of useful properties that arbitrary power series do not share, so they deserve being given a name. If $(a_n)_{n=0}^{\infty}$ is a sequence in $\mathbb{K}$ such that $a_n \neq 0$ for at most finitely many indices n then the power series

$$a(x) = \sum_{n=0}^{\infty} a_n x^n$$

is called a *polynomial* (over $\mathbb{K}$). The maximum index d with $a_d \neq 0$ is called the *degree* of the polynomial $a(x)$ and denoted by $\deg a(x)$. For the zero power series, which has no nonzero coefficients at all, we define $\deg 0 := -\infty$. For a nonzero polynomial $a(x)$, the coefficient $[x^{\deg a(x)}]a(x)$ is called the *leading coefficient* of $a(x)$, and it is denoted by $\operatorname{lc} a(x)$. We leave the leading coefficient of 0 undefined.

M. Kauers, P. Paule, *The Concrete Tetrahedron*
© Springer-Verlag/Wien 2011

The sum of two polynomials is again a polynomial, and so is the product of two polynomials. Therefore, the polynomials form a subring of $\mathbb{K}[[x]]$, denoted by $\mathbb{K}[x]$. The ring of polynomials inherits many properties from the ring of power series in the obvious way, but there are some differences as well.

On the one hand, some operations are allowed for general power series but not for polynomials. For example, we have seen that every power series $a(x)$ with $a(0) \neq 0$ has a multiplicative inverse. So in particular, every polynomial $a(x)$ with $a(0) \neq 0$ has a multiplicative inverse. However, this inverse is in general not a polynomial anymore, but a power series with infinitely many nontrivial terms (think of the inverse of $1 - x$). The polynomials and their multiplicative inverses form a field

$$\mathbb{K}(x) := \{p(x)/q(x) : p(x), q(x) \in \mathbb{K}[x], q(x) \neq 0\} \subsetneq \mathbb{K}((x)).$$

Its elements are traditionally called *rational functions*, although, just like generating functions, they are algebraic objects rather than actual functions.

On the other hand, some operations are allowed for polynomials which cannot be performed for arbitrary power series. For example, we have seen that in order for the composition $a(b(x))$ of two power series to be meaningful, we need in general that $b(0) = 0$. This condition is not needed when $a(x)$ is a polynomial, because $b(x)^k$ is meaningful for every fixed $k \in \mathbb{N}$, and if $a(x)$ is a polynomial, then $a(b(x))$ is just a finite linear combination of some $b(x)^k$.

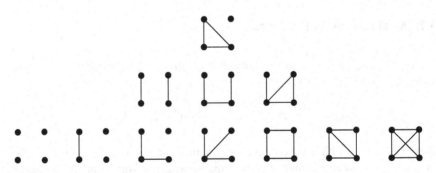

Fig. 3.1 The 11 unlabeled graphs with four vertices

In particular, the composition $a(b(x))$ is defined when $a(x)$ is a polynomial and $b(x)$ is just a constant, i.e., if $a(x) = \sum_{n=0}^{d} a_n x^n \in \mathbb{K}[x]$ and $b(x) = b_0 \in \mathbb{K}$. We then have

$$a(b_0) = a_0 + a_1 b_0 + a_2 b_0^2 + \cdots + a_d b_0^d \in \mathbb{K}$$

and call this value the *evaluation* of the polynomial $a(x)$ at $x = b_0$. Via the evaluation, every polynomial, although at first introduced as a formal object only, admits an interpretation as an actual function $\mathbb{K} \to \mathbb{K}$.

Both $\mathbb{K}[[x]]$ and $\mathbb{K}[x]$ share the property of being a vector space over $\mathbb{K}$, but unlike for $\mathbb{K}[[x]]$, a *basis* of $\mathbb{K}[x]$ can be easily given: each polynomial is a finite linear combination of the monomials $1, x, x^2, x^3$, etc. As the monomials are clearly linearly independent over $\mathbb{K}$, they form a basis. We call it the *standard basis* of $\mathbb{K}[x]$. This is the most natural basis, but it is not in all circumstances the most convenient basis to work with, and it is generally not a bad idea to switch to alternative bases whenever this seems to bring an advantage. For example, the *falling factorial basis* is advantageous for summation problems, as we shall see later. This basis consists of the terms $1, x, x^{\underline{2}}, x^{\underline{3}}, x^{\underline{4}}, \ldots$, where

$$x^{\underline{n}} := x(x-1)(x-2)\cdots(x-n+1) \qquad (n \geq 0)$$

is called the n-th falling factorial of x.

We can easily express the $x^{\underline{n}}$ in terms of the standard basis by simply expanding the product, giving

$$x^{\underline{0}} = 1,$$
$$x^{\underline{1}} = x,$$
$$x^{\underline{2}} = x^2 - x,$$
$$x^{\underline{3}} = x^3 - 3x^2 + 2x,$$
$$x^{\underline{4}} = x^4 - 6x^3 + 11x^2 - 6x,$$
$$x^{\underline{5}} = x^5 - 10x^4 + 35x^3 - 50x^2 + 24x, \quad \text{etc.},$$

and since the $x^{\underline{n}}$ form a basis, it is also possible to express the standard monomials in terms of this basis:

$$1 = x^{\underline{0}},$$
$$x = x^{\underline{1}},$$
$$x^2 = x^{\underline{2}} + x^{\underline{1}},$$
$$x^3 = x^{\underline{3}} + 3x^{\underline{2}} + x^{\underline{1}},$$
$$x^4 = x^{\underline{4}} + 6x^{\underline{3}} + 7x^{\underline{2}} + x^{\underline{1}},$$
$$x^5 = x^{\underline{5}} + 10x^{\underline{4}} + 25x^{\underline{3}} + 15x^{\underline{2}} + x^{\underline{1}}, \quad \text{etc.}$$

The coefficients arising in converting one of these bases to the other are called the *Stirling numbers* of the first and the second kind, respectively, and denoted by $S_1(n,k)$ and $S_2(n,k)$, respectively. Precisely, these numbers are defined through the formulas

$$x^{\underline{n}} = \sum_{k=0}^{n} S_1(n,k)x^k \quad \text{and} \quad x^n = \sum_{k=0}^{n} S_2(n,k)x^{\underline{k}} \qquad (n \geq 0).$$

Closely related to the falling factorials are the *rising factorials*. They are written and defined as

$$x^{\overline{n}} := x(x+1)(x+2)\cdots(x+n-1) \qquad (n \geq 0),$$

and they also form a basis of $\mathbb{K}[x]$ (Problem 3.5). Traditionally, rising factorials are also denoted by the Pochhammer symbol $(x)_n := x^{\overline{n}}$.

Polynomials are inherently finite objects; they are easily represented on a computer by their finite list of coefficients with respect to a fixed (infinite) basis. The computational treatment of polynomials is at the heart of computer algebra, and every computer algebra system provides a large number of procedures applicable to polynomials. We refer to the standard text books on computer algebra [22, 58] for the corresponding algorithmic background.

3.2 Polynomials as Sequences

Polynomials can be evaluated at arbitrary points. For a fixed polynomial $p(x) \in \mathbb{K}[x]$ of $\deg p(x) = d$, we can therefore consider in particular the sequence

$$p(0), \ p(1), \ p(2), \ p(3), \ p(4), \ \ldots$$

obtained by evaluating the polynomial at all the natural numbers. Such sequences are called *polynomial sequences* of degree d (not to be confused with sequences of polynomials). For example,

$$1, \ 4, \ 9, \ 16, \ 25, \ 36, \ \ldots, \ n^2, \ \ldots$$

is a polynomial sequence of degree 2.

The sum of two polynomial sequences is obviously again a polynomial sequence. Also the Hadamard product of two polynomial sequences is clearly again a polynomial sequence. What about the Cauchy product? That is, if $p(x)$ and $q(x)$ are polynomials, does then the convolution

$$\sum_{k=0}^{n} p(k)q(n-k)$$

necessarily agree with the evaluations $u(n)$ of some polynomial $u(x) \in \mathbb{K}[x]$? To see that this is indeed the case, let us determine the generating function of a polynomial sequence.

We have already seen in Chap. 1 that generating functions for some polynomial sequences can be obtained by differentiating the geometric series:

$$\frac{1}{1-x} = \sum_{n=0}^{\infty} 1 x^n,$$

$$D_x \frac{1}{1-x} = \frac{1}{(1-x)^2} = \sum_{n=1}^{\infty} n x^{n-1} = \sum_{n=0}^{\infty} (n+1)x^n,$$

$$D_x^2 \frac{1}{1-x} = 2 \frac{1}{(1-x)^3} = \sum_{n=2}^{\infty} n(n-1)x^{n-2} = \sum_{n=0}^{\infty} (n+1)(n+2)x^n,$$

$$D_x^3 \frac{1}{1-x} = 6 \frac{1}{(1-x)^4} = \sum_{n=3}^{\infty} n(n-1)(n-2)x^{n-3} = \sum_{n=0}^{\infty} (n+1)(n+2)(n+3)x^n,$$

$$\vdots$$

An easy induction confirms that

$$\frac{k!}{(1-x)^{k+1}} = \sum_{n=0}^{\infty} (n+1)^{\overline{k}} x^n \qquad (k \geq 0),$$

and since every polynomial can be written in terms of rising factorials, this is already enough to cover the general case. For example, to get the generating function of $(n^5)_{n=0}^{\infty}$, write

$$n^5 = -(n+1)^{\overline{0}} + 31(n+1)^{\overline{1}} - 90(n+1)^{\overline{2}} + 65(n+1)^{\overline{3}} - 15(n+1)^{\overline{4}} + (n+1)^{\overline{5}}$$

and find

$$\sum_{n=0}^{\infty} n^5 x^n = -\frac{1}{1-x} + \frac{31 \cdot 1}{(1-x)^2} - \frac{90 \cdot 2}{(1-x)^3} + \frac{65 \cdot 6}{(1-x)^4} - \frac{15 \cdot 24}{(1-x)^5} + \frac{120}{(1-x)^6}$$

$$= \frac{x^5 + 26x^4 + 66x^3 + 26x^2 + x}{(x-1)^6}.$$

In general, if $p(x)$ is a polynomial of degree d, then we will have

$$\sum_{n=0}^{\infty} p(n)x^n = \frac{q(x)}{(1-x)^{d+1}}$$

for some polynomial $q(x)$ of degree at most d.

For the converse, suppose that $q(x)$ is some polynomial of degree at most d. Then $q(x)$ can be written by repeated polynomial division with remainder by $(1-x)$ in the form

$$q(x) = q_0 + q_1(1-x) + q_2(1-x)^2 + \cdots + q_d(1-x)^d$$

for some $q_0, \ldots, q_d \in \mathbb{K}$. Dividing the equation by $(1-x)^{d+1}$ shows that

$$\frac{q(x)}{(1-x)^{d+1}} = \frac{q_0}{d!} \frac{d!}{(1-x)^{d+1}} + \frac{q_1}{(d-1)!} \frac{(d-1)!}{(1-x)^d} + \cdots + \frac{q_d}{1} \frac{1}{1-x},$$

and therefore, $q(x)/(1-x)^{d+1}$ is the generating function of a polynomial sequence. Putting things together, we obtain the following characterization result.

Theorem 3.1 *A power series $a(x) \in \mathbb{K}[[x]]$ is the generating function of a polynomial sequence of degree $d \geq 0$ if and only if $a(x) = q(x)/(1-x)^{d+1}$ for some polynomial $q(x) \in \mathbb{K}[x]$ of degree at most d with $q(1) \neq 0$.*

Now it is also clear that the Cauchy product of two polynomial sequences is again a polynomial sequence, for the simple reason that

$$\frac{q_1(x)}{(1-x)^{d_1}} \frac{q_2(x)}{(1-x)^{d_2}} = \frac{q_1(x)q_2(x)}{(1-x)^{d_1+d_2}}$$

and $\deg q_1(x)q_2(x) = \deg q_1(x) + \deg q_2(x)$.

3.3 The Tetrahedron for Polynomials

The most accessible vertex of the Concrete Tetrahedron for polynomial sequences is the asymptotic estimation: the asymptotics of a polynomial sequence is just the maximum degree term, and there is nothing more to say about it. More interesting are the other vertices and the relations between them.

The *forward difference* of a sequence $(a_n)_{n=0}^{\infty}$ is defined as the sequence $(a_{n+1} - a_n)_{n=0}^{\infty}$. We write Δ for the operator that maps a sequence to its forward difference. Consider a sequence $(a_n)_{n=0}^{\infty}$ and let $b_n := \Delta a_n = a_{n+1} - a_n$. It is an easy matter to express the generating function $b(x)$ of $(b_n)_{n=0}^{\infty}$ in terms of the generating function $a(x)$ of $(a_n)_{n=0}^{\infty}$:

$$b_n = a_{n+1} - a_n \qquad (n \geq 0)$$

implies

$$b(x) = \sum_{n=0}^{\infty} b_n x^n = \sum_{n=0}^{\infty} (a_{n+1} - a_n)x^n = \frac{1}{x} \sum_{n=1}^{\infty} a_n x^n - \sum_{n=0}^{\infty} a_n x^n$$

$$= \frac{1}{x}(a(x) - a(0)) - a(x) = \frac{(1-x)a(x) - a(0)}{x}.$$

If $(a_n)_{n=0}^{\infty}$ is a polynomial sequence of degree d, then $a(x) = q(x)/(1-x)^{d+1}$ for some polynomial $q(x)$ of degree less than $d + 1$, by Theorem 3.1. It follows that

$$b(x) = \frac{(1-x)a(x) - a(0)}{x} = \frac{(q(x) - q(x)(1-x)^d)/x}{(1-x)^d}.$$

This means that the forward difference operator reduces the degree of a polynomial sequence by one. Applying the operator $d + 1$ times in a row, we will obtain the zero sequence:

$$\Delta^{d+1} a_n = 0.$$

This translates into simple recurrence equations for polynomial sequences. For example, the sequence $a_n = n^2$ (as well as any other polynomial sequence of degree two) satisfies the recurrence

$$\Delta^3 a_n = \Delta^2 (a_{n+1} - a_n) = \Delta (a_{n+2} - 2a_{n+1} + a_n)$$

$$= a_{n+3} - 3a_{n+2} + 3a_{n+1} - a_n = 0 \qquad (n \geq 0).$$

The degree drop upon applying the Delta operator to a polynomial sequence parallels the degree drop upon applying the derivative to a polynomial. More specifically, we have the analogue

$$\Delta n^{\underline{d}} = d n^{\underline{d-1}} \qquad \longleftrightarrow \qquad D_x x^d = d x^{d-1}.$$

In the same vein as forward difference parallels differentiation, summation of sequences parallels integration of power series. For example, we have the straightforward analogues

$$\sum_{k=0}^{n} \Delta u_k = u_{n+1} - u_0 \quad \longleftrightarrow \quad \int_x D_t u(t)\,dt = u(x) - u(0),$$

$$\Delta \sum_{k=0}^{n} u_k = u_{n+1} \quad \longleftrightarrow \quad D_x \int_x u(t)\,dt = u(x),$$

$$\sum_{k=0}^{n} u_k \Delta v_k = u_{n+1}v_{n+1} - u_0 v_0 \quad \longleftrightarrow \quad \int_x u(t) D_t v(t)\,dt = u(x)v(x) - u(0)v(0)$$

$$- \sum_{k=0}^{n} (\Delta u_k) v_{k+1} \qquad\qquad - \int_x (D_t u(t)) v(t)\,dt$$

and several others. As far as the summation of polynomials is concerned, these observations together imply the summation formula

$$\sum_{k=0}^{n} k^{\underline{d}} = \frac{1}{d+1}(n+1)^{\underline{d+1}} \quad \longleftrightarrow \quad \int_x t^d\,dt = \frac{1}{d+1} t^{d+1},$$

which gives rise to a simple algorithm for summing polynomial sequences: Given a polynomial $p(x) \in \mathbb{K}[x]$, we can compute a polynomial $q(x) \in \mathbb{K}[x]$ such that $q(n) = \sum_{k=0}^{n} p(k)$ $(n \geq 0)$ by simply writing $p(x)$ as a linear combination of falling factorials and then applying the above formula term by term. For example, in order to evaluate $\sum_{k=0}^{n} k^5$, write

$$k^5 = \sum_{i=0}^{5} S_2(5,i) k^{\underline{i}} = k^{\underline{1}} + 15 k^{\underline{2}} + 25 k^{\underline{3}} + 10 k^{\underline{4}} + k^{\underline{5}}$$

and conclude that

$$\sum_{k=0}^{n} k^5 = \sum_{k=0}^{n} k^{\underline{1}} + 15 \sum_{k=0}^{n} k^{\underline{2}} + 25 \sum_{k=0}^{n} k^{\underline{3}} + 10 \sum_{k=0}^{n} k^{\underline{4}} + \sum_{k=0}^{n} k^{\underline{5}}$$

$$= \tfrac{1}{2}(n+1)^{\underline{2}} + \tfrac{15}{3}(n+1)^{\underline{3}} + \tfrac{25}{4}(n+1)^{\underline{4}} + \tfrac{10}{5}(n+1)^{\underline{5}} + \tfrac{1}{6}(n+1)^{\underline{6}}$$

$$= \tfrac{1}{12}(2n^6 + 6n^5 + 5n^4 - n^2).$$

A closed form representation for every sum over a polynomial sequence can be found in this way, and it is just a polynomial sequence whose degree is one higher. That such a closed form always exists does not come as a surprise if we look at the generating functions. Namely, if $a(x)$ is the generating function of some sequence $(a_n)_{n=0}^{\infty}$, then

$$\frac{1}{1-x} a(x) = \sum_{n=0}^{\infty} \left(\sum_{k=0}^{n} a_k \right) x^n$$

is the generating function for the sequence of its partial sums, and by Theorem 3.1 this is the generating function of a polynomial sequence of degree $d+1$ whenever $a(x)$ is the generating function of a polynomial sequence of degree d. In fact, the summation algorithm just described is essentially equivalent to switching from a given polynomial to its generating function, multiplying by $1/(1-x)$, and switching back to the coefficient sequence.

There is another way of evaluating the sum over a polynomial sequence in closed form. Given $p(x) \in \mathbb{K}[x]$ of degree d, we know that the desired closed form of $\sum_{k=0}^{n} p(k)$ will be a polynomial of degree $d+1$. Since a polynomial of degree $d+1$ is uniquely determined by its values at $d+2$ different points, we can just evaluate the sum for $d+2$ different specific values in place of n and compute the interpolating polynomial of these values. In the example $p(x) = x^5$, the sum $\sum_{k=0}^{n} p(k)$ evaluates to

$$0,\ 1,\ 33,\ 276,\ 1300,\ 4425,\ 12201$$

for $n = 0, \ldots, 6$, and there will be exactly one polynomial of degree 6 matching those values. We can find this polynomial either by using an interpolation algorithm or by making a brute-force ansatz

$$q(x) = q_0 + q_1 x + q_2 x^2 + q_3 x^3 + q_4 x^4 + q_5 x^5 + q_6 x^6$$

with undetermined coefficients $q_0, \ldots, q_6$ and solving the linear system of equations that is obtained by equating the linear form $q(n)$ to the respective values of the sum for $n = 0, \ldots, 6$. Either way will give us the polynomial $q(x) = \frac{1}{12}(2x^6 + 6x^5 + 5x^4 - x^2)$ we have already found before.

3.4 Polynomials as Solutions

Evaluating a sum $s_n := \sum_{k=0}^{n} p(k)$ in closed form is equivalent to solving the telescoping equation

$$p(n+1) = s_{n+1} - s_n \qquad (n \geq 0)$$

in closed form. Being able to sum polynomials, we therefore also know how to solve such equations in terms of polynomials whenever $p(n)$ is a polynomial sequence. In fact, there is nothing specific to sequences in this equation, we may as well consider it as a purely algebraic problem in $\mathbb{K}[x]$: Given $p(x) \in \mathbb{K}[x]$, we can find $a(x) \in \mathbb{K}[x]$ such that

$$p(x) = a(x+1) - a(x).$$

What about more complicated equations? Let us consider equations of the form

$$p(x) = q(x)a(x+1) - r(x)a(x),$$

where $p(x), q(x), r(x) \in \mathbb{K}[x]$ are polynomials and we seek polynomial solutions $a(x) \in \mathbb{K}[x]$ to the equation. A polynomial solution $a(x)$ for an equation of this form need not exist, as is easily seen from the choice $p(x) = 1$, $q(x) = 0$, $r(x) = x$. What we are after is an algorithm that either finds a polynomial solution, or proves that no such solution exists.

The problem is easily solved if we restrict the attention to polynomial solutions $a(x)$ of fixed degree d. In this case, we can make an ansatz

$$a(x) = a_0 + a_1 x + a_2 x^2 + \cdots + a_d x^d$$

with undetermined coefficients $a_0, \dots, a_d \in \mathbb{K}$, plug this form into the equation and compare coefficients of like powers of x. This leads to a linear system of equations for the a_i whose solution space consists precisely of the coefficient vectors of all the desired polynomial solutions $a(x)$. For example, in order to find a quadratic polynomial $a(x)$ satisfying the equation

$$3x^2 - 3 = 2x\,a(x+1) - (2x+1)a(x),$$

we plug $a(x) = a_0 + a_1 x + a_2 x^2$ into the equation and compare coefficients with respect to x, obtaining the equation system

$$-3 = -a_0, \quad 0 = -a_1 - 2a_2, \quad 3 = 3a_2$$

whose unique solution is $(a_0, a_1, a_2) = (3, -2, 1)$. Therefore, $a(x) = x^2 - 2x + 3$ is the only polynomial solution of degree at most two that satisfies the equation we considered.

If there is no restriction on the degree, we can make a blind guess, say $d = 500$, and proceed as above. If we are lucky, we find a solution. But if we find no solution, then we cannot be sure whether no solution exists or our guess for the degree was just too small. In order to solve the general problem, we need to analyze how high a degree of a polynomial solution can possibly be. Such an analysis is a little technical, but it can be done. There are several cases to distinguish. The easy case is when $\deg q(x) \neq \deg r(x)$ or $\mathrm{lc}\,q(x) \neq \mathrm{lc}\,r(x)$. Then, since $\deg a(x+1) = \deg a(x) =: d$ and $\mathrm{lc}\,a(x+1) = \mathrm{lc}\,a(x)$ for a hypothetical solution $a(x)$, the leading terms on the right hand side of our equation

$$p(x) = q(x)a(x+1) - r(x)a(x)$$

cannot cancel each other, so they must be canceled by the leading term of $p(x)$. This implies that $d = \deg p(x) - \max(\deg q(x), \deg r(x))$ for every potential polynomial solution $a(x)$ of the equation. The case where a cancellation can happen is more tricky. Suppose that $\deg q(x) = \deg r(x)$ and $\mathrm{lc}\,q(x) = \mathrm{lc}\,r(x)$. Then the degree of the right hand side is less than $d + \deg q(x)$, because the highest degree term drops out. The key idea is now to rewrite $a(x+1)$ as $(a(x+1) - a(x)) + a(x)$, so that the equation becomes

$$p(x) = q(x)(a(x+1) - a(x)) - (r(x) - q(x))a(x).$$

Since $\deg(a(x+1)-a(x))=d-1$, the degree of $q(x)(a(x+1)-a(x))$ is exactly one less than the degree of $q(x)a(x)$, and by assumption, the degree of $r(x)-q(x)$ is (one or more) less than the degree of $q(x)$, because the leading terms of $r(x)$ and $q(x)$ cancel. If more terms in $r(x)-q(x)$ cancel, that is if the degree drops by more than one, then the leading term of the right hand side is determined by $q(x)(a(x+1)-a(x))$ and it must be canceled by the leading term of $p(x)$, so $d=\deg p(x)-\deg q(x)+1$. Otherwise, if $\deg(r(x)-q(x))=\deg q(x)-1$, there can again be a cancellation between the leading terms of $q(x)(a(x+1)-a(x))$ and $(r(x)-q(x))a(x)$. In that case, the leading coefficients must also match. Because of $\mathrm{lc}(a(x+1)-a(x))=d\,\mathrm{lc}\,a(x)$, this situation requires

$$d\,\mathrm{lc}\,q(x)\,\mathrm{lc}\,a(x)=\mathrm{lc}(r(x)-q(x))\,\mathrm{lc}\,a(x)$$

and thus forces $d=\mathrm{lc}(r(x)-q(x))/\mathrm{lc}\,q(x)$. In all cases, we have found a finite bound on the degree of a potential polynomial solution $a(x)$ that depends just on the given polynomials $p(x),q(x),r(x)$:

- If $\deg q(x)\neq\deg r(x)$ or $\mathrm{lc}\,q(x)\neq\mathrm{lc}\,r(x)$, then $d=\deg p(x)-\max(\deg q(x),\deg r(x))$.
- If $\deg q(x)=\deg r(x)$ and $\deg(r(x)-q(x))<\deg q(x)-1$, then $d=\deg p(x)-\deg q(x)+1$.
- If $\deg q(x)=\deg r(x)$ and $\deg(r(x)-q(x))=\deg q(x)-1$ and $\frac{\mathrm{lc}(r(x)-q(x))}{\mathrm{lc}\,q(x)}$ is not an integer, then $d=\deg p(x)-\deg q(x)+1$.
- If $\deg q(x)=\deg r(x)$ and $\deg(r(x)-q(x))=\deg q(x)-1$ and $\frac{\mathrm{lc}(r(x)-q(x))}{\mathrm{lc}\,q(x)}$ is an integer, then $d\leq\max(\deg p(x)-\deg q(x)+1,\mathrm{lc}(r(x)-q(x))/\mathrm{lc}\,q(x))$.

The equation

$$p(x)=q(x)a(x+1)-r(x)a(x)$$

can thus be solved by first picking from the above case distinction an appropriate number d that bounds the degree of a possible solution $a(x)\in\mathbb{K}[x]$. With this degree bound at hand, we can make an ansatz with undetermined coefficients for $a(x)$, plug it into the equation, compare coefficients, and solve a linear system. Any solution of the linear system gives a solution $a(x)$ of the equation, and if the linear system has no solution, then there is no polynomial $a(x)$ that satisfies the equation.

In the example from before, where $p(x)=3x^2-3$, $q(x)=2x$, $r(x)=2x+1$, we have $\deg q(x)=\deg r(x)$ and $\deg(r(x)-q(x))=0=\deg q(x)-1$ and $\mathrm{lc}(r(x)-q(x))/\mathrm{lc}\,q(x)=1/2$ is not an integer, therefore we are in the third case and expect

$$\deg a(x)=\deg p(x)-\deg q(x)+1=2-1+1=2$$

for any possible solution $a(x)$. Indeed, this prediction matches the degree of the solution we found.

As an example for the fourth case, consider the equation

$$15 = x^2 a(x+1) - (x^2 + 5x + 1)a(x).$$

With $p(x) = 15$, $q(x) = x^2$ and $r(x) = x^2 + 5x + 1$, we have $\deg q(x) = \deg r(x) = 2$, $\deg(r(x) - q(x)) = 1 = \deg q(x) - 1$, and $\mathrm{lc}(r(x) - q(x))/\mathrm{lc}\,q(x) = 5/1 = 5$ is an integer. On the other hand, $\deg p(x) - \deg q(x) + 1 = 0 - 1 + 1 = 0 < 5$, so if the equation has a polynomial solution at all, then this solution will have degree at most five. Plugging a generic quintic polynomial $a(x) = a_0 + a_1 x + \cdots + a_5 x^5$ into the equation and comparing coefficients with respect to x leads to a linear system for the undetermined coefficients a_i. As this system happens to have a solution, our equation does posses a polynomial solution. Here it is:

$$a(x) = 8x^5 + 72x^4 + 220x^3 + 256x^2 + 75x - 15.$$

If the linear system had been unsolvable, we could have concluded that there was no polynomial solution (of any degree).

3.5 Polynomials as Coefficients

Sometimes polynomials themselves appear as elements of a sequence. The most simple sequence of polynomials is the sequence $(x^n)_{n=0}^{\infty}$ of monomials, other simple examples are $(x^{\underline{n}})_{n=0}^{\infty}$ or $(x^{\overline{n}})_{n=0}^{\infty}$. More interesting sequences of polynomials appear as coefficients in certain formal power series expansions. For example, the power series

$$\frac{y \exp(xy)}{\exp(y) - 1} = 1 + \tfrac{1}{2}(2x-1)y + \tfrac{1}{12}(6x^2 - 6x + 1)y^2 + \tfrac{1}{12}(2x^3 - 3x^2 + x)y^3$$
$$+ \tfrac{1}{720}(30x^4 - 60x^3 + 30x^2 - 1)y^4 + \cdots$$

lives not only in $\mathbb{K}[[x,y]]$ but even in $\mathbb{K}[x][[y]]$, as this is obviously the case for $\exp(xy)$ and $y/(\exp(y) - 1)$ and $\mathbb{K}[x][[y]]$ is closed under multiplication. The polynomial $B_n(x) := n![y^n]y\exp(xy)/(\exp(y) - 1)$ is known as the n-th *Bernoulli polynomial*. It reduces to the Bernoulli numbers for $x = 0$, and one of its numerous interesting properties is that it gives a simple expression for monomial sums (Problem 3.9):

$$\sum_{k=1}^{n} k^d = \frac{1}{d+1}\left(B_{d+1}(n+1) - B_{d+1}(0)\right).$$

Another interesting sequence of polynomials appears in the series expansion

$$\frac{1 - xy}{1 - 2xy + y^2} = 1 + xy + (2x^2 - 1)y^2 + (4x^3 - 3x)y^3 + (8x^4 - 8x^2 + 1)y^4 + \cdots.$$

As $[y^0](1-2xy+y^2) = 1$ is invertible in $\mathbb{K}$, the denominator polynomial $1-2xy+y^2$ has a multiplicative inverse in $\mathbb{K}[x][[y]]$, and consequently also $(1-xy)/(1-2xy+y^2)$ belongs to $\mathbb{K}[x][[y]]$. The polynomials $T_n(x) := [y^n](1-xy)/(1-2xy+y^2)$ are called *Chebyshev polynomials of the first kind,* and we will have a closer look at them in the next chapter. One of their key features is that they form a vector space basis of $\mathbb{R}[x]$ which is orthogonal in the sense that

$$\int_{-1}^{1} \frac{1}{\sqrt{1-t^2}} T_n(t) T_m(t)\, dt = 0 \quad\Longleftrightarrow\quad n \neq m.$$

More generally, for any fixed integrable function $w\colon (a,b) \to (0,\infty)$ we can define a scalar product $\langle .,.\rangle_w$ on the space of integrable functions $f,g\colon (a,b) \to \mathbb{R}$ via

$$\langle f,g\rangle_w := \int_a^b w(t)f(t)g(t)\, dx.$$

A sequence $(p_n(x))_{n=0}^{\infty}$ of polynomials in $\mathbb{R}[x]$ with $\deg p_n(x) = n$ $(n \geq 0)$ is then called a sequence of *orthogonal polynomials* if it is orthogonal with respect to one such scalar product $\langle .,.\rangle_w$ in the sense that

$$\langle p_n, p_m\rangle_w = 0 \quad\Longleftrightarrow\quad n \neq m.$$

Sequences of orthogonal polynomials are important for example in numerical analysis, because they can be used to construct sparse linear systems in situations where the standard monomial basis would only produce dense systems.

Power series with polynomial coefficients are also useful in combinatorics. Let, for example, $a_{n,k}$ be the number of unlabeled graphs with n vertices and k edges. Then $a_n(x) = \sum_{k=0}^{\infty} a_{n,k} x^k$ is a polynomial for every fixed $n \in \mathbb{N}$, because no graph with n vertices can have more than $\frac{1}{2}n(n-1)$ edges. (We have already met $a_4(x)$ at the beginning of this chapter.) Therefore,

$$\begin{aligned}
a(x,y) := \sum_{n=0}^{\infty} a_n(x)y^n &= 1+y+(1+x)y^2+(1+x+x^2+x^3)y^3 \\
&\quad + (1+x+2x^2+3x^3+2x^4+x^5+x^6)y^4 \\
&\quad + (1+x+2x^2+4x^3+6x^4+6x^5+6x^6+4x^7+2x^8+x^9+x^{10})y^5 + \cdots
\end{aligned}$$

belongs to $\mathbb{K}[x][[y]]$. Quite some information about unlabeled graphs is implicitly contained in this series. For example:

- Evaluation at $x = 1$ (note that we are allowed to do this) gives the generating function for the number of graphs with n vertices and arbitrary number of edges:

$$a(1,y) = 1+y+2y^2+4y^3+11y^4+34y^5+\cdots$$

- The average number of edges in a graph with n vertices is found to be

$$\frac{[y^n]D_x a(x,y)|_{x=1}}{[y^n]a(1,y)} = \frac{\sum_{k=0}^{\infty} k a_{n,k}}{\sum_{k=0}^{\infty} a_{n,k}}.$$

- If we allow multiple edges between two vertices, then the number of graphs with n vertices and k edges is given by the coefficient of $x^k y^n$ in $a(\frac{x}{1-x}, y) \in \mathbb{K}[[x,y]]$.
- If we have two different kinds of vertices, then the number of graphs with k edges, n_1 vertices of the first kind and n_2 vertices of the second kind is given by the coefficient of $x^k y_1^{n_1} y_2^{n_2}$ in the power series $a(x, y_1 + y_2) \in \mathbb{K}[x][[y_1, y_2]]$.

In short, the series $a(x,y)$ refines the counting of the series $a(1,y)$ in that it distinguishes objects of the same size (number of vertices) according to some other property (number of edges). To drive this idea to its extreme, we can let the objects themselves appear in the coefficient sequence. For example, the series

$$\frac{1}{(1-x_1 y)(1-x_2 y^2)(1-x_3 y^3)} = 1 + x_1 y + (x_1^2 + x_2) y^2 + (x_1^3 + x_1 x_2 + x_3) y^3$$
$$+ (x_1^4 + x_1^2 x_2 + x_2^2 + x_1 x_3) y^4 + (x_1^5 + x_1^3 x_2 + x_1 x_2^2 + x_1^2 x_3 + x_2 x_3) y^5$$
$$+ (x_1^6 + x_1^4 x_2 + x_1^2 x_2^2 + x_2^3 + x_1^3 x_3 + x_1 x_2 x_3 + x_3^2) y^6 + \cdots$$

generates all the ways to write an integer n as a sum of 1s, 2s, and 3s: each possible representation $n = a \cdot 1 + b \cdot 2 + c \cdot 3$ is encoded by a monomial $x_1^a x_2^b x_3^c$ that appears in the coefficient polynomial of y^n. This explains the origin of the word "generating function". For emphasis, a power series that generates all the combinatorial objects from a certain class is sometimes also called the *complete* generating function.

3.6 Applications

Polynomials have applications in virtually every branch of mathematics. Also polynomial sequences arise in numerous different contexts. We restrict to three examples.

Figurate Numbers

Let us start with something simple. We have seen in the introduction that the simple sum $\sum_{k=1}^{n} k$ can be illustrated pictorially by an appropriate arrangement of pebbles. For this reason, the numbers $T_n := \sum_{k=1}^{n} k = \frac{1}{2} n(n+1)$ are also known as the *triangular numbers*. What about other geometric figures instead of triangles? Not too surprisingly, n^2 pebbles are needed to fill a square. For a pentagon, consider the illustration in Fig. 3.2. In order to complete a pentagon with side length n to a pentagon of side length $n+1$, we need to place $3n+1$ new pebbles around the old one. This gives the recurrence

$$P_{n+1} = P_n + (3n+1) \qquad (n \geq 1),$$

which together with the initial value $P_1 = 1$ uniquely determines all the pentagonal numbers P_n. As a closed form we find $P_n = \frac{1}{2} n(3n-1)$.

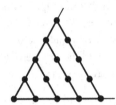 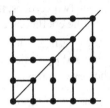 

Fig. 3.2 Triangular numbers, square numbers, and pentagonal numbers

In general, for a regular k-gon ($k \geq 3$), the same reasoning leads to the recurrence

$$P_{n+1}^{(k)} = P_n^{(k)} + (k-2)n + 1 \qquad (n \geq 1), \qquad P_1^{(k)} = 1,$$

which admits the closed form $P_n^{(k)} = \frac{1}{2}n(kn - 2n - k + 4)$.

Graph Colorings

Another context in which polynomial sequences arise is the graph coloring problem. A (labeled) graph is a pair $G = (V, E)$, where V is a non-empty finite set and E is a set of 2-element subsets of V. The elements of V are the *vertices* of the graph, and we say that two vertices $v_1, v_2 \in V$ are connected by an *edge* if $\{v_1, v_2\} \in E$. Our task is to assign colors to the vertices of a given graph in such a way that no two vertices connected by an edge bear the same color.

Whether such a coloring exists depends on how many colors we have available. If we have as many colors as the graph has vertices, then we can easily solve the problem by assigning a different color to each vertex. Whether less than $n := |V|$ colors suffice, this depends on the structure of the graph. The question we are interested in is the number of ways in which a fixed graph G can be colored by k colors or less, where k varies.

Let $P(G, k)$ be this number. If c_i denotes the number of ways to color G with *exactly* i different colors, then $c_i = 0$ for all $i > n$. Having k colors available, we can form $c_i \binom{k}{i}$ different colorings where exactly i of the k colors are used. From this we obtain

$$P(G, k) = \sum_{i=1}^{\infty} c_i \binom{k}{i} = \sum_{i=1}^{n} c_i \binom{k}{i} \qquad (k \geq 1)$$

for the total number of colorings of G with k colors or less. Since $n = |V|$ is constant and

$$\binom{k}{i} = \frac{1}{i!}k(k-1)(k-2)\cdots(k-i+1),$$

the numbers $P(G, k)$ form a polynomial sequence in k. It is called the *chromatic polynomial* of G.

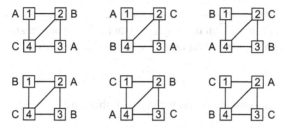

Fig. 3.3 All six 3-colorings of a graph with four vertices

Consider as an example the graph depicted in Fig. 3.3. For this graph, we have $c_1 = c_2 = 0$ (by inspection), $c_3 = 6$ (by the picture), $c_4 = 4! = 24$ (by lack of restrictions), and $c_i = 0$ for $i > 4$. Therefore,

$$P(G,k) = 6\binom{k}{3} + 24\binom{k}{4} = k^4 - 5k^3 + 8k^2 - 4k \quad (k \geq 1)$$

in this case.

Partition Analysis

In additive number theory, a partition of a non-negative integer n is a representation of n as a sum of non-negative integers. The form of this representation may be restricted further by some constraints. When no constraints are imposed, a partition is simply a sum of positive integers in which the order of the summands does not matter. These partitions are counted by the partition number p_n which we already introduced in Chap. 2. The numbers p_n have quite a complicated nature, but surprisingly, counting more complicated arrangements may lead to less complicated counting sequences. As an example, for some fixed $k \in \mathbb{N}$ consider matrices of size $k \times k$ with entries in $\mathbb{N}$ which are such that all rows and all columns sum up to a prescribed number $m \in \mathbb{N}$. Such matrices are called *magic squares* of size k with *magic constant m*. For example,

6	4	5	5
0	9	8	3
2	6	5	7
12	1	2	5

is a magic square of size $k = 4$ with magic constant $m = 20$.

How does the number of magic squares depend on the magic constant when we fix the size? For $k = 1$ and $k = 2$ it is easily checked that there are precisely 1 and $m + 1$ squares, respectively. From $k > 2$ on the counting becomes less trivial, but it can be shown in general that the number of magic squares of size k is a polynomial sequence of degree $(k+1)^2$ in the magic constant m. Without making an attempt at proving anything here, we just report that when $k = 3$, there are

$$\tfrac{1}{8}(m+1)(m+2)(m^2+3m+4)$$

squares with magic constant m, and when $k = 4$, there are

$$\tfrac{1}{11340}(m+1)(m+2)(m+3)(11m^6+132m^5+683m^4+1944m^3 \\ +3320m^2+3360m+1890).$$

Unlike for figurate numbers where there is a general expression having both m and k as variables, no general expression is known for counting magic squares. However, an algorithm is known which can compute the generating function for the counting sequence for any specific choice of k.

There is a part of partition theory, called *partition analysis,* which provides algorithms that are applicable not only to magic squares but to a more general type of integer partitions that can be specified in terms of systems of linear equations and inequalities. Another example for this theory is to count the number of non-congruent triangles with prescribed perimeter $n \in \mathbb{N} \setminus \{0\}$ and sides a,b,c of positive integer length. Rephrased in terms of inequalities, we are looking for the number of tuples (a,b,c) satisfying the conditions $n = a+b+c$, $1 \le a \le b \le c$, $a \ge b+c$, $b \ge a+c$, and $c \ge a+b$. For $n = 9$, there are three such tuples: $(3,3,3)$, $(2,3,4)$, and $(1,4,4)$. In general, if T_n denotes the number of such tuples, then the generating function is given by

$$\sum_{n=0}^{\infty} T_n x^n = \frac{x^3}{(1-x^2)(1-x^3)(1-x^4)} = x^3+x^5+x^6+2x^7+x^8+3x^9+\cdots.$$

Taking this result for granted, it is clear that $(T_n)_{n=0}^{\infty}$ is not a polynomial sequence, because the denominator is not of the form $(1-x)^m$. Still, all the roots of the denominator live on the unit circle of the complex plane. Coefficient sequences of such power series are called *quasi polynomials.* We will not discuss them further, because they are included as a special case in the class of C-finite sequences discussed in the following chapter.

3.7 Problems

Problem 3.1 Is there a difference between a polynomial and a truncated power series?

Problem 3.2 Is there a difference between $\mathbb{K}[x][[y]]$ and $\mathbb{K}[[y]][x]$?

Problem 3.3 Is there a difference between $\mathbb{K}[x]$ and the ring $\mathrm{Pol}(\mathbb{K})$ of polynomial functions, which is defined as

$$\mathrm{Pol}(\mathbb{K}) := \{ f \in \mathbb{K}^{\mathbb{K}} : \exists \, p \in \mathbb{K}[x] \; \forall \, x \in \mathbb{K} : f(x) = p(x) \}$$

with pointwise addition and multiplication?

Problem 3.4 Prove the following polynomial version of Theorem 2.6: For every fixed $u(x) \in \mathbb{K}[x]$ the map

$$\Phi_u : \mathbb{K}[x] \to \mathbb{K}[x], \qquad a(x) \mapsto a(u(x))$$

is a ring homomorphism. (Note that this polynomial version includes the case $u(x) = u_0 \in \mathbb{K}$, which is excluded in Theorem 2.6. In this special case, Φ_u is called *evaluation homomorphism*.)

Problem 3.5 Use the relation $x^{\overline{n}} = (-1)^n (-x)^{\underline{n}}$ ($n \in \mathbb{N}$) to deduce formulas for converting polynomials from the standard basis to the rising factorial basis and back.

Problem 3.6 Convert the following polynomials into the falling factorial basis.

1. $x^3 + 2x^2 + 2x + 1,$ 2. $2x^3 + 5x^2 - 4x + 2,$ 3. $5x^3 + 2x^2 + 3.$

Problem 3.7 Show that the Stirling numbers satisfy

$$S_1(n,k) = (1-n)S_1(n-1,k) + S_1(n-1,k-1) \quad (n,k > 0),$$
$$S_2(n,k) = kS_2(n-1,k) + S_2(n-1,k-1) \qquad (n,k > 0).$$

Deduce the basis conversion formulas

$$x^{\underline{n}} = \sum_{k=0}^{n} S_1(n,k) x^k \quad \text{and} \quad x^n = \sum_{k=0}^{n} S_2(n,k) x^{\underline{k}} \qquad (n \ge 0)$$

from these recurrence equations and the initial conditions.

Problem 3.8 Show $\sum_{n=k}^{\infty} \frac{S_2(n,k)}{n!} x^n = \frac{1}{k!}(e^x - 1)^k$ and use this to prove the identity $B_n = \sum_{k=0}^{n} S_2(n,k)$, where B_n denotes the n-th Bell number.

Problem 3.9 Let $B_n(x) := n! [y^n] \frac{y \exp(xy)}{\exp(y)-1}$ be the Bernoulli polynomials and $B_n = B_n(0)$ be the Bernoulli numbers.

1. Prove the identity

$$\sum_{d=0}^{\infty} \left(\sum_{k=0}^{n} k^d \right) \frac{y^d}{d!} = \sum_{k=0}^{n} \exp(ky) \qquad (n \ge 0)$$

and use it to derive the summation formula

$$\sum_{k=0}^{n} k^d = \frac{1}{d+1} \left(B_{d+1}(n+1) - B_{d+1} \right) \qquad (n,d \ge 0).$$

2. Prove the identity

$$B_n(x) = \sum_{k=0}^{n} \binom{n}{k} B_k x^{n-k} \qquad (n \geq 0)$$

and use it to rewrite the summation formula from part 1 in the form

$$\sum_{k=0}^{n} k^d = \frac{1}{d+1} \sum_{k=0}^{d} B_k \binom{d+1}{k} (n+1)^{d-k+1} \qquad (n, d \geq 0).$$

(Observe that the sum on the right is an explicit polynomial in n whenever d is a specific integer, so this formula provides yet another way for summing polynomial sequences. This formula is due to Jacob Bernoulli.)

Problem 3.10 Prove that $(H_n)_{n=0}^{\infty}$ is not a polynomial sequence.

Problem 3.11 In a computer algebra system of your choice, write a program that takes as input a polynomial $p \in \mathbb{Q}[x]$ and returns as output the generating function $\sum_{n=0}^{\infty} p(n) x^n$ as a quotient of two polynomials.

Problem 3.12 Let $(u_n)_{n=0}^{\infty}$ and $(v_n)_{n=0}^{\infty}$ be sequences. Prove that the difference operator satisfies the product rule

$$\Delta(u_k v_k) = (\Delta u_k) v_k + u_{k+1} \Delta v_k \qquad (k \geq 0)$$

and use this law to deduce the formula for summation by parts:

$$\sum_{k=0}^{n} u_k \Delta v_k = u_{n+1} v_{n+1} - u_0 v_0 - \sum_{k=0}^{n} (\Delta u_k) v_{k+1} \qquad (n \geq 0).$$

Problem 3.13 For each of the following equations, find a polynomial solution or prove that there is none.

1. $27 = (2x^2 + 1)a(x+1) - (2x^2 + 8x + 3)a(x),$
2. $2x^2 - 6x + 3 = (2x^2 + 8x + 1)a(x+1) - (2x^2 + 8x + 3)a(x),$
3. $2x^2 + 6x + 3 = (2x^2 + 8x + 1)a(x+1) - (2x^2 + 8x + 3)a(x).$

Check whether your favorite computer algebra system provides a command for solving such equations.

Problem 3.14 Given $p(x), q(x), r(x) \in \mathbb{K}[x]$, deduce a bound on the possible degree of a polynomial $a(x) \in \mathbb{K}[x]$ satisfying the differential equation

$$p(x) = q(x) D_x a(x) - r(x) a(x).$$

Problem 3.15 Extend the degree analysis of Sect. 3.4 to inhomogeneous equations of second order. That is, given $p(x), q(x), r(x), s(x) \in \mathbb{K}[x]$, deduce a bound on the possible degree of a polynomial $a(x) \in \mathbb{K}[x]$ satisfying the recurrence equation

$$p(x) = q(x)a(x+2) + r(x)a(x+1) + s(x)a(x).$$

Problem 3.16 How many ways are there to color the following graph with 1000 colors?

Chapter 4
C-Finite Sequences

Polynomial sequences obey certain linear recurrence equations with constant coefficients, as we have seen in the previous chapter. But in general, a sequence satisfying a linear recurrence equation with constant coefficients need not be a polynomial sequence. The solutions to such recurrence equations form a strictly larger class of sequences, the so-called C-finite sequences. This is the class we consider now.

4.1 Fibonacci Numbers

Fibonacci numbers are classically introduced as the total number of offspring a single pair of rabbits produces in n months, assuming that the initial pair is born in the first month, and that every pair which is two months or older produces another pair of rabbits per month (Fig. 4.1). The growth of the population under these assumptions is easily described: In the n-th month, we have all the pairs of rabbits that have already been there in the $(n-1)$-st month (rabbits don't die in our model), and all the new pairs of rabbits born in the n-th month. The number of pairs born in the n-th month is just the number of pairs who were already there in the $(n-2)$-nd month. The number of pairs of rabbits in the n-th month is therefore given by the n-th Fibonacci number, which is recursively defined via

$$F_{n+2} = F_{n+1} + F_n \quad (n \geq 0), \qquad F_0 = 0, F_1 = 1.$$

The first terms of this sequence are

$$0, \ 1, \ 1, \ 2, \ 3, \ 5, \ 8, \ 13, \ 21, \ 34, \ 55, \ 89, \ 144, \ldots$$

Although this sequence has been studied at least one thousand years before it was popularized in the west in 1202 by Leonardo da Pisa, it still today attracts the attention of a large number of both professional and amateur mathematicians. This is partly because of the simple definition it has, partly because of the multitude of non-

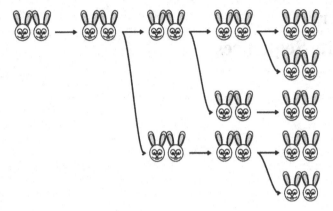

Fig. 4.1 Multiplying pairs of rabbits

mathematical contexts in which it arises, and partly because of some pretty peculiar mathematical properties it enjoys.

Lots of facts about Fibonacci numbers are available in classical textbooks on the subject or even on the web. We do not have the place here to give even a narrow overview over the material, and will only make some introductory remarks. The reader is encouraged to type "Fibonacci numbers" into a search engine and explore further properties on his or her own.

The recurrence equation quoted above allows us to compute F_n for every n recursively, by first computing F_{n-1} and F_{n-2} and then adding the two results to obtain F_n. If this is programmed naively, it leads to a horrible performance: If T_n is the number of additions needed to compute F_n in this way, we have $T_n = T_{n-1} + T_{n-2} + 1$ ($n \geq 2$), $T_0 = T_1 = 0$. A quick induction confirms that $T_n > F_n$ for $n \geq 2$, so computing

$$F_{100} = 354224848179261915075$$

in this way would take more than 10000 years even under the favorable assumption that we can do 10^9 additions per second. There must be a better way.

The problem is that there is an unnecessary amount of recomputation going on. For computing F_n, the naive program will first compute F_{n-1} recursively, then compute F_{n-2} recursively, and then return their sum. A clever implementation will take into account that F_{n-2} was already computed during the computation of F_{n-1}, and does not need to be computed again. The most easy way to avoid recomputation is to turn the second order recurrence into a first order matrix recurrence:

$$\begin{pmatrix} F_{n+1} \\ F_{n+2} \end{pmatrix} = \begin{pmatrix} 0 & 1 \\ 1 & 1 \end{pmatrix} \begin{pmatrix} F_n \\ F_{n+1} \end{pmatrix} \qquad (n \geq 0).$$

By unfolding this recurrence in the obvious way, we can determine F_{n+1} and F_{n+2} simultaneously using n additions only. The computation of F_{100} is now a piece of cake.

But this is not the end. The unwinding of the recurrence is nothing else than a repeated multiplication of a fixed matrix to the initial values. Written in the form

$$\begin{pmatrix} F_n \\ F_{n+1} \end{pmatrix} = \begin{pmatrix} 0 & 1 \\ 1 & 1 \end{pmatrix}^n \begin{pmatrix} F_0 \\ F_1 \end{pmatrix} \qquad (n \geq 0),$$

we see that computing the n-th Fibonacci number is really not more than raising a certain matrix to the n-th power. Taking into account that $F_0 = 0$ and $F_1 = 1$, and setting $F_{-1} = 1$ (which is consistent with the recurrence), we can write this also as

$$\begin{pmatrix} F_{n-1} & F_n \\ F_n & F_{n+1} \end{pmatrix} = \begin{pmatrix} 0 & 1 \\ 1 & 1 \end{pmatrix}^n \begin{pmatrix} F_{-1} & F_0 \\ F_0 & F_1 \end{pmatrix} = \begin{pmatrix} 0 & 1 \\ 1 & 1 \end{pmatrix}^n \qquad (n \geq 0).$$

An immediate consequence of this equation is the relation

$$\begin{pmatrix} F_{2n-1} & F_{2n} \\ F_{2n} & F_{2n+1} \end{pmatrix} = \begin{pmatrix} 0 & 1 \\ 1 & 1 \end{pmatrix}^{2n} = \begin{pmatrix} 0 & 1 \\ 1 & 1 \end{pmatrix}^n \begin{pmatrix} 0 & 1 \\ 1 & 1 \end{pmatrix}^n = \begin{pmatrix} F_{n-1} & F_n \\ F_n & F_{n+1} \end{pmatrix}^2 \qquad (n \geq 0),$$

which implies

$$F_{2n} = F_n(F_{n+1} + F_{n-1}) = F_n(2F_{n+1} - F_n)$$
$$\text{and } F_{2n+1} = F_n^2 + F_{n+1}^2$$

for $n \geq 0$. These formulas break any pair (F_n, F_{n+1}) down to (F_0, F_1) in just $\log_2(n)$ iterations. A single iteration requires two additions and four multiplications, making together $6\log_2(n)$ operations. For computing F_{100}, this scheme needs 40 operations, which may not seem like a big improvement compared to the 100 additions we needed before, but the difference becomes more and more significant as n grows: it certainly makes a difference whether $F_{1000000000}$ is computed with 180 operations or with 1000000000. On the other hand, counting only the number of operations is not really fair, because the computation time depends also on the length of the integers appearing as intermediate results. If these are properly taken into account, it turns out that both schemes have a linear runtime. But in finite domains, where numbers have a fixed size, the advantage of the logarithmic algorithm is striking.

The representation of Fibonacci numbers as a matrix power is not only of computational interest. It is also the source of several important identities. For example, taking determinants on both sides of

$$\begin{pmatrix} F_{n-1} & F_n \\ F_n & F_{n+1} \end{pmatrix} = \begin{pmatrix} 0 & 1 \\ 1 & 1 \end{pmatrix}^n \qquad (n \geq 0)$$

gives directly Cassini's identity

$$F_{n-1}F_{n+1} - F_n^2 = (-1)^n \qquad (n \geq 0).$$

Another famous identity follows from diagonalization. The decomposition

$$\begin{pmatrix} 0 & 1 \\ 1 & 1 \end{pmatrix} = \underbrace{\begin{pmatrix} 1 & 1 \\ \phi_1 & \phi_2 \end{pmatrix}}_{=:T} \underbrace{\begin{pmatrix} \phi_1 & 0 \\ 0 & \phi_2 \end{pmatrix}}_{=:D} \underbrace{\begin{pmatrix} 1 & 1 \\ \phi_1 & \phi_2 \end{pmatrix}^{-1}}_{=T^{-1}}$$

with $\phi_1 = \frac{1}{2}(1 + \sqrt{5})$ and $\phi_2 = \frac{1}{2}(1 - \sqrt{5})$ implies

$$\begin{pmatrix} F_{n-1} & F_n \\ F_n & F_{n+1} \end{pmatrix} = (TDT^{-1})^n = TD^nT^{-1}$$

$$= \begin{pmatrix} 1 & 1 \\ \phi_1 & \phi_2 \end{pmatrix} \begin{pmatrix} \phi_1^n & 0 \\ 0 & \phi_2^n \end{pmatrix} \begin{pmatrix} 1 & 1 \\ \phi_1 & \phi_2 \end{pmatrix}^{-1} \qquad (n \geq 0).$$

Carrying out the matrix product leads to the Euler-Binet formula

$$F_n = \frac{1}{\sqrt{5}} \left(\left(\frac{1 + \sqrt{5}}{2} \right)^n - \left(\frac{1 - \sqrt{5}}{2} \right)^n \right) \qquad (n \geq 0).$$

The number $\phi = \frac{1}{2}(1 + \sqrt{5}) \approx 1.61803$ appearing here is known as the *golden ratio*. Because of $|\frac{1}{2}(1 - \sqrt{5})| \approx |-0.61803| < \phi$, the term ϕ^n dominates the asymptotic behavior. We have

$$F_n \sim \frac{1}{\sqrt{5}} \phi^n \qquad (n \to \infty),$$

the error being so small that F_n is actually equal to the integer closest to $\frac{1}{\sqrt{5}} \phi^n$ for every $n \geq 0$. Another consequence of the asymptotic estimate is

$$\lim_{n \to \infty} \frac{F_{n+1}}{F_n} = \phi.$$

4.2 Recurrences with Constant Coefficients

The Fibonacci numbers are a prototypical example for a sequence that satisfies a linear recurrence with constant coefficients. Such equations are called *C-finite recurrences*, and a sequence $(a_n)_{n=0}^{\infty}$ is called C-finite if it satisfies a C-finite equation. To be explicit, a sequence $(a_n)_{n=0}^{\infty}$ is called C-finite (of order r) if there are numbers $c_0, c_1, \ldots, c_{r-1} \in \mathbb{K}$ with $c_0 \neq 0$ such that

$$a_{n+r} + c_{r-1}a_{n+r-1} + \cdots + c_1 a_{n+1} + c_0 a_n = 0 \qquad (n \geq 0).$$

All the terms in a C-finite sequence of order r are uniquely determined by the coefficients $c_0, \ldots, c_{r-1}$ of the recurrence equation and the initial values $a_0, a_1, \ldots, a_{r-1}$. It is important to note that this is only a finite amount of data and can be stored and manipulated faithfully in a computer algebra system. (Hence the name C-finite, the C refers to the constant coefficients.)

For a fixed C-finite recurrence, different initial values lead to different sequences. The set of all sequences that satisfy some fixed C-finite recurrence equation forms a vector space over $\mathbb{K}$, for if two sequences $(a_n)_{n=0}^{\infty}$ and $(b_n)_{n=0}^{\infty}$ are such that

$$a_{n+r} + c_{r-1}a_{n+r-1} + \cdots + c_1 a_{n+1} + c_0 a_n = 0 \qquad (n \geq 0)$$
$$\text{and} \quad b_{n+r} + c_{r-1}b_{n+r-1} + \cdots + c_1 b_{n+1} + c_0 b_n = 0 \qquad (n \geq 0),$$

then multiplying the first equation by $\alpha \in \mathbb{K}$, the second by $\beta \in \mathbb{K}$, and adding them together gives

$$(\alpha a_{n+r} + \beta b_{n+r}) + \cdots + c_1(\alpha a_{n+1} + \beta b_{n+1}) + c_0(\alpha a_n + \beta b_n) = 0 \qquad (n \geq 0),$$

so the linear combination $(\alpha a_n + \beta b_n)_{n=0}^\infty$ satisfies the same recurrence as $(a_n)_{n=0}^\infty$ and $(b_n)_{n=0}^\infty$. The vector space of all the sequences that satisfy some fixed C-finite recurrence is called the *solution space* of that recurrence.

The solution space of a C-finite recurrence has only a finite dimension, in fact, its dimension equals the order r of the recurrence. To see that it has at least this dimension, it is sufficient to guarantee that there are r linearly independent solutions. This is easy because any choice of the r initial values can be extended in a unique way to a solution of the recurrence, so if for $i = 0, \ldots, r-1$ we let $(b_n^{(i)})_{n=0}^\infty$ be the solutions that start like

$$0, 0, \ldots, 0, \underset{\underset{\text{index } i}{\uparrow}}{1}, 0, \ldots, 0, b_r^{(i)}, b_{r+1}^{(i)}, b_{r+2}^{(i)}, \ldots,$$

then these r sequences are clearly linearly independent over $\mathbb{K}$. This proves that the solution space has at least dimension r. To prove that its dimension cannot be more than r, it is sufficient to observe that every solution is uniquely determined by its first r values, all the further values being forced by the recurrence. So if $(a_n)_{n=0}^\infty$ is any solution to the recurrence, then

$$a_n = a_0 b_n^{(0)} + a_1 b_n^{(1)} + \cdots + a_{r-1} b_n^{(r-1)} \qquad (n \geq 0),$$

because both sides agree for $n = 0, \ldots, r-1$ and both sides are solutions of the recurrence. In short, what we have shown is that the operation of truncating a sequence $(a_n)_{n=0}^\infty$ to a vector $(a_n)_{n=0}^{r-1} \in \mathbb{K}^r$ induces a vector space isomorphism between the solution space of the recurrence and $\mathbb{K}^r$.

With the solutions $(b_n^{(i)})_{n=0}^\infty$ $(i = 0, \ldots, r-1)$, we already know a vector space basis of the solution space. But these solutions are not particularly handy. Our next goal is to construct a basis that consists of sequences which admit closed form expressions. We have already seen in the previous section that the Fibonacci numbers can be expressed as a linear combinations of the two *exponential sequences* $(\phi^n)_{n=0}^\infty$ and $((-\phi)^{-n})_{n=0}^\infty$ where $\phi = \frac{1}{2}(1 + \sqrt{5})$ is the golden ratio. Indeed, these two sequences themselves form a basis of the solution space of the Fibonacci recurrence. Does this generalize to arbitrary C-finite sequences? Let's see. Suppose that the recurrence

$$a_{n+r} + c_{r-1} a_{n+r-1} + \cdots + c_1 a_{n+1} + c_0 a_n = 0 \qquad (n \geq 0)$$

has a solution $(u^n)_{n=0}^\infty$ for some $u \in \mathbb{K} \setminus \{0\}$. Then

$$u^{n+r} + c_{r-1} u^{n+r-1} + \cdots + c_1 u^{n+1} + c_0 u^n = 0 \qquad (n \geq 0),$$

and dividing both sides by u^n leads to

$$u^r + c_{r-1} u^{r-1} + \cdots + c_1 u + c_0 = 0,$$

a plain polynomial equation for u which no longer depends on n. The polynomial $x^r + c_{r-1}x^{r-1} + \cdots + c_1 x + c_0 \in \mathbb{K}[x]$ is called the *characteristic polynomial* of the recurrence. Every root u of the characteristic polynomial gives rise to a solution $(u^n)_{n=0}^\infty$ of the corresponding recurrence. For example, in the case of the Fibonacci recurrence

$$a_{n+2} - a_{n+1} - a_n = 0,$$

the characteristic polynomial is $x^2 - x - 1$, and its two roots are $\frac{1}{2}(1 + \sqrt{5})$ and $\frac{1}{2}(1 - \sqrt{5})$, in accordance with what we have found before.

The characteristic polynomial suggests the viewpoint of operators acting on sequences. If we define the mapping

$$\bullet : \mathbb{K}[x] \times \mathbb{K}^{\mathbb{N}} \to \mathbb{K}^{\mathbb{N}}$$
$$(c_r x^r + c_{r-1} x^{r-1} + \cdots + c_0) \bullet (a_n)_{n=0}^\infty := (c_r a_{n+r} + \cdots + c_0 a_n)_{n=0}^\infty,$$

then a C-finite recurrence can be stated compactly in the form $c(x) \bullet (a_n)_{n=0}^\infty = 0$ where $c(x) \in \mathbb{K}[x]$ is the characteristic polynomial of the recurrence. In this setting, monomials x^i can be regarded as shift operators $n \mapsto n + i$. Polynomial arithmetic is compatible with the composition of operators in the sense that

$$(u(x) + v(x)) \bullet (a_n)_{n=0}^\infty = u(x) \bullet (a_n)_{n=0}^\infty + v(x) \bullet (a_n)_{n=0}^\infty$$
$$\text{and } (u(x)v(x)) \bullet (a_n)_{n=0}^\infty = u(x) \bullet (v(x) \bullet (a_n)_{n=0}^\infty)$$

for every $u(x), v(x) \in \mathbb{K}[x]$ and every sequence $(a_n)_{n=0}^\infty$. As a consequence, if $(a_n)_{n=0}^\infty$ is a solution of $v(x) \bullet (a_n)_{n=0}^\infty = 0$ for some $v(x) \in \mathbb{K}[x]$, then also $(u(x)v(x)) \bullet (a_n)_{n=0}^\infty = 0$, because

$$(u(x)v(x)) \bullet (a_n)_{n=0}^\infty = u(x) \bullet (v(x) \bullet (a_n)_{n=0}^\infty) = u(x) \bullet 0 = 0.$$

In terms of operators, finding exponential solutions $(u^n)_{n=0}^\infty$ of a C-finite recurrence amounts to finding linear factors $x - u$ of the characteristic polynomial, because $(u^n)_{n=0}^\infty$ is a solution of the recurrence $a_{n+1} - u a_n = 0$ which corresponds to the linear polynomial $x - u$.

If $\mathbb{K}$ is algebraically closed, then we know that a characteristic polynomial of degree r will split into r linear factors. Each linear factor gives rise to a solution of the recurrence, and if all the linear factors are different, then we have found r different solutions of the recurrence. These are linearly independent (Problem 4.7) and so they form a basis of the solution space. It remains to consider what happens when the characteristic polynomial has multiple roots. We have actually seen characteristic polynomials with multiple roots already in the previous chapter, where we observed that any polynomial sequence $(p(n))_{n=0}^\infty$ with $p(x) \in \mathbb{K}[x]$ of degree 2 satisfies the recurrence

$$\Delta^3 p(n) = p(n+3) - 3p(n+2) + 3p(n+1) - p(n) = 0 \qquad (n \geq 0).$$

The characteristic polynomial of this recurrence is $(x - 1)^3$. The observation is that multiple roots of the characteristic polynomial induce polynomial factors in the solutions. In the following theorem, we give the general result.

Theorem 4.1 *Suppose that $c_0, \ldots, c_{r-1} \in \mathbb{K}$ with $c_0 \neq 0$ are such that*

$$x^r + c_{r-1} x^{r-1} + \cdots + c_1 x + c_0 = (x - u_1)^{e_1} (x - u_2)^{e_2} \cdots (x - u_m)^{e_m}$$

for $e_1, \ldots, e_m \in \mathbb{N} \setminus \{0\}$ and pairwise distinct $u_1, \ldots, u_m \in \mathbb{K}$. Then the sequences $(n^i u_j^n)_{n=0}^{\infty}$ $(j = 1, \ldots, m, \ i = 0, \ldots, e_j - 1)$ form a basis of the $\mathbb{K}$-vector space of all solutions of the recurrence equation

$$a_{n+r} + c_{r-1} a_{n+r-1} + \cdots + c_1 a_{n+1} + c_0 a_n = 0 \qquad (n \geq 0).$$

Proof. There are two things to show: (*i*) all the sequences $(n^i u_j^n)_{n=0}^{\infty}$ are solutions of the recurrence, and (*ii*) all other solutions are linear combinations of those.

(*i*) Take an arbitrary $j \in \{1, \ldots, m\}$ and $i \in \{0, \ldots, e_j - 1\}$. We show that $(n^i u_j^n)_{n=0}^{\infty}$ is a solution of the recurrence. As $(x - u_j)^{i+1}$ is a factor of the characteristic polynomial, it suffices to show that $(x - u_j)^{i+1} \bullet (n^i u_j^n)_{n=0}^{\infty} = 0$, which is easily done by induction on i.

(*ii*) Since the characteristic polynomial has degree r, we have $e_1 + \cdots + e_m = r$, so the number of solutions we are considering agrees with the dimension of the solution space. What remains to be shown is that the solutions stated in the theorem are linearly independent. Suppose this is not the case. Then there is a nontrivial $\mathbb{K}$-linear combination of them which is identically zero. By grouping sequences of like exponential parts together, we can write this linear combination in the form

$$p_1(n) u_1^n + p_2(n) u_2^n + \cdots + p_m(n) u_m^n = 0 \qquad (n \geq 0),$$

with certain $p_1(x), \ldots, p_m(x) \in \mathbb{K}[x]$ not all of which are zero. Our plan is to take such a linear combination which is minimal in a certain sense, and then construct a smaller one. This gives the desired contradiction.

Although not all of $p_1(x), \ldots, p_m(x) \in \mathbb{K}[x]$ are zero, some of them may be. For every relation, there will be an index i such that $p_1(x) = \cdots = p_{i-1}(x) = 0 \neq p_i(x)$. From all the assumed relations, select those where this index i is as large as possible. Among those, we may consider one in which $\deg p_i(x)$ is as small as possible. Then we have

$$p_i(n) u_i^n + p_{i+1}(n) u_{i+1}^n + \cdots + p_m(n) u_m^n = 0 \qquad (n \geq 0).$$

This equation implies both

$$p_i(n+1) u_i^{n+1} + p_{i+1}(n+1) u_{i+1}^{n+1} + \cdots + p_m(n+1) u_m^{n+1} = 0 \quad \text{(shift } n \mapsto n+1)$$

and

$$p_i(n) u_i^{n+1} + p_{i+1}(n) u_i u_{i+1}^n + \cdots + p_m(n) u_i u_m^n = 0 \qquad \text{(multiply by } u_i).$$

Subtracting the two equations gives

$$(\Delta p_i)(n) u_i u_i^n + \tilde{p}_{i+1}(n) u_{i+1}^n + \cdots + \tilde{p}_m(n) u_m^n = 0 \qquad (n \geq 0),$$

an equation that violates our minimality assumptions, for either $\Delta p_i(x) = 0$, then i was not maximal, or $\Delta p_i(x) \neq 0$, then $\deg \Delta p_i(x) = \deg p_i(x) - 1$ and so $\deg p_i(x)$ was not minimal. $\qquad \square$

This theorem allows us to express any C-finite sequence as a linear combination of terms of the form $n^d u^n$. All we need to do is to determine the roots of the characteristic polynomials along with their multiplicities and then find a linear combination of the terms predicted in the theorem above that matches the initial values of the sequence at hand. For example, for the sequence $(a_n)_{n=0}^{\infty}$ defined via

$$a_{n+4} - 4a_{n+3} + 2a_{n+2} + 4a_{n+1} - 3a_n = 0 \qquad (n \geq 0)$$
$$\text{and } a_0 = 1, \quad a_1 = 3, \quad a_2 = -4, \quad a_3 = 0,$$

we find first that

$$x^4 - 4x^3 + 2x^2 + 4x - 3 = (x-1)^2(x+1)(x-3).$$

By the theorem there must be some constants c_1, c_2, c_3, c_4 such that

$$a_n = c_1 + c_2 n + c_3(-1)^n + c_4 3^n \qquad (n \geq 0).$$

These constants are quickly determined by setting $n = 0, 1, 2, 3, 4$, which gives a linear system for the c_i. It is a consequence of Theorem 4.1 that this system must have a unique solution. In our case, we find $c_1 = \frac{13}{4}$, $c_2 = -3$, $c_3 = -\frac{19}{8}$, $c_4 = \frac{1}{8}$, so the final result is

$$a_n = \frac{13}{4} - 3n - \frac{19}{8}(-1)^n + \frac{1}{8}3^n \qquad (n \geq 0).$$

4.3 Closure Properties

There are C-finite sequences that arise from some specific combinatorial or number theoretic considerations. The most famous one is certainly the Fibonacci sequence. Another one is the sequence of *Lucas numbers* $(L_n)_{n=0}^{\infty}$ defined via

$$L_{n+2} - L_{n+1} - L_n = 0 \quad (n \geq 0), \qquad L_0 = 2, L_1 = 1.$$

This sequence behaves in many ways similarly to the Fibonacci numbers.

Another C-finite sequence is the Perrin sequence $(P_n)_{n=0}^{\infty}$, which is defined by

$$P_{n+3} - P_{n+1} - P_n = 0 \quad (n \geq 0), \qquad P_0 = 3, P_1 = 0, P_2 = 2.$$

The first numbers in this sequence are

$$3, \ 0, \ 2, \ 3, \ 2, \ 5, \ 5, \ 7, \ 10, \ 12, \ 17, \ldots$$

Perrin numbers are known for a curious number theoretic property they have. It appears that $n \in \mathbb{N}$ is prime "almost if and only if" $P_n \bmod n = 0$. For small n, we have

n	2	3	4	5	6	7	8	9	10	11	12
$P_n \bmod n$	0	0	2	0	5	0	2	3	7	0	5

The first mismatch is at $n = 271441$, for which P_n mod $n = 0$ although $271441 = 521^2$ is composite. On the other hand, it can be shown that if n is prime, then definitely P_n mod $n = 0$, and therefore, the Perrin numbers provide a strong necessary condition for a natural number to be a prime.

We know from the previous section that the Fibonacci and the Lucas numbers satisfy the same linear recurrence, so any linear combination of them will satisfy that recurrence as well. In particular, we find that the numbers $F_n + L_n$ are C-finite. What about the numbers $F_n + P_n$? Do they also form a C-finite sequence? Yes, they do, and it is not hard to come up with a recurrence satisfied by them. We just need to take the recurrence whose characteristic polynomial is the product of the two respective characteristic polynomials for F_n and P_n. This recurrence will have $(F_n)_{n=0}^{\infty}$ as a solution, because its characteristic polynomial is a multiple of $x^2 - x - 1$, and it will have $(P_n)_{n=0}^{\infty}$ as a solution, because its characteristic polynomial is a multiple of $x^3 - x - 1$. Consequently, also $(F_n + P_n)_{n=0}^{\infty}$ will be a solution of this recurrence.

By the same reasoning, it follows that whenever $(a_n)_{n=0}^{\infty}$ and $(b_n)_{n=0}^{\infty}$ are C-finite sequences then so is their sum $(a_n + b_n)_{n=0}^{\infty}$. We say that the class of C-finite sequences is *closed under addition*. There are several other closure properties by which new C-finite sequences can be obtained from known ones. The following theorem provides an overview.

Theorem 4.2 *Let $(u_n)_{n=0}^{\infty}$ and $(v_n)_{n=0}^{\infty}$ be C-finite sequences in $\mathbb{K}$ of order r and s, respectively, and let $m \in \mathbb{N}$, $m \geq 1$. Then:*

1. $(u_n + v_n)_{n=0}^{\infty}$ *is C-finite of order at most $r + s$,*
2. $(u_n v_n)_{n=0}^{\infty}$ *is C-finite of order at most rs,*
3. $(\sum_{k=0}^{n} u_k)_{n=0}^{\infty}$ *is C-finite of order at most $r + 1$,*
4. $(u_{mn})_{n=0}^{\infty}$ *is C-finite of order at most r,*
5. $(u_{\lfloor n/m \rfloor})_{n=0}^{\infty}$ *is C-finite of order at most mr.*

Proof. If $c_0, \ldots, c_{r-1} \in \mathbb{K}$ are such that

$$u_{n+r} + c_{r-1} u_{n+r-1} + \cdots + c_1 u_{n+1} + c_0 u_n = 0 \qquad (n \geq 0),$$

then also

$$u_{n+i} + c_{r-1} u_{n+i-1} + \cdots + c_1 u_{n+i-r+1} + c_0 u_{n+i-r} = 0 \qquad (n \geq 0)$$

for every fixed $i \geq r$. Repeated use of the recurrence shows that $(u_{n+i})_{n=0}^{\infty}$ ($i \geq 0$ fixed) is a linear combination of $(u_n)_{n=0}^{\infty}, (u_{n+1})_{n=0}^{\infty}, \ldots, (u_{n+r-1})_{n=0}^{\infty}$. In other words, the shifted sequences $(u_{n+i})_{n=0}^{\infty}$ ($i \geq 0$ fixed) all belong to the $\mathbb{K}$-vector space U of dimension at most r generated by the sequences $(u_n)_{n=0}^{\infty}, \ldots, (u_{n+r-1})_{n=0}^{\infty}$. Analogously, the sequences $(v_{n+i})_{n=0}^{\infty}$ ($i \geq 0$ fixed) all belong to the $\mathbb{K}$-vector space V of dimension (at most) s generated by the sequences $(v_n)_{n=0}^{\infty}, \ldots, (v_{n+s-1})_{n=0}^{\infty}$.

1. Let $(w_n)_{n=0}^{\infty} = (u_n)_{n=0}^{\infty} + (v_n)_{n=0}^{\infty}$. All shifted sequences $(w_{n+i})_{n=0}^{\infty}$ ($i \geq 0$ fixed) belong to the $\mathbb{K}$-vector space $W := U + V$ generated by $(u_n)_{n=0}^{\infty}, \ldots, (u_{n+r-1})_{n=0}^{\infty}$ and $(v_n)_{n=0}^{\infty}, \ldots, (v_{n+s-1})_{n=0}^{\infty}$, because this vector space contains all the shifted sequences $(u_{n+i})_{n=0}^{\infty}$ and all the shifted sequences $(v_{n+i})_{n=0}^{\infty}$. The dimension of W is at most $r + s$, so any $r + s + 1$ sequences $(w_{n+i})_{n=0}^{\infty}$ ($i \geq 0$ fixed) must be linearly dependent. In particular, the sequences

$$(w_n)_{n=0}^{\infty}, \ (w_{n+1})_{n=0}^{\infty}, \ \ldots, (w_{n+r+s})_{n=0}^{\infty} \in W$$

must be linearly dependent over $\mathbb{K}$. This means there are constants $d_0, \ldots, d_{r+s} \in \mathbb{K}$, not all zero, such that

$$d_0 w_n + d_1 w_{n+1} + \cdots + d_{r+s} w_{n+r+s} = 0 \qquad (n \geq 0),$$

and therefore $(w_n)_{n=0}^{\infty}$ is C-finite of order at most $r + s$.

2. Let $(w_n)_{n=0}^{\infty} = (u_n)_{n=0}^{\infty} \odot (v_n)_{n=0}^{\infty} = (u_n v_n)_{n=0}^{\infty}$. All shifted sequences $(w_{n+i})_{n=0}^{\infty}$ ($i \geq 0$ fixed) belong to the $\mathbb{K}$-vector space $W = U \otimes V$ generated by the mutual products

$$(u_n v_n)_{n=0}^{\infty}, (u_{n+1} v_n)_{n=0}^{\infty}, \ldots, (u_{n+r-1} v_n)_{n=0}^{\infty},$$

$$(u_n v_{n+1})_{n=0}^{\infty}, (u_{n+1} v_{n+1})_{n=0}^{\infty}, \ldots, (u_{n+r-1} v_{n+1})_{n=0}^{\infty},$$

$$\vdots$$

$$(u_n v_{n+s-1})_{n=0}^{\infty}, (u_{n+1} v_{n+s-1})_{n=0}^{\infty}, \ldots, (u_{n+r-1} v_{n+s-1})_{n=0}^{\infty}.$$

Arguing as before, since W has dimension at most rs, the sequence $(w_n)_{n=0}^{\infty}$ must satisfy a recurrence with constant coefficients of order at most rs.

3. Now let $(w_n)_{n=0}^{\infty} = (\sum_{k=0}^{n} u_k)_{n=0}^{\infty}$. Because of

$$\sum_{k=0}^{n+i} u_k = \left(\sum_{k=0}^{n} u_k\right) + u_{n+1} + u_{n+2} + \cdots + u_{n+i} \qquad (n \geq 0)$$

for every fixed $i \geq 0$, all the shifted sequences $(w_{n+i})_{n=0}^{\infty}$ ($i \geq 0$ fixed) belong to the vector space W generated by $(w_n)_{n=0}^{\infty}$ and $(u_n)_{n=0}^{\infty}, \ldots, (u_{n+r-1})_{n=0}^{\infty}$. Since W has dimension at most $r + 1$, the claim follows.

4. and 5. Problem 4.5.   □

Theorem 4.2 is constructive in the sense that recurrences for the sequences proved C-finite can be obtained computationally from recurrences for of the given sequences by making the linear algebra reasoning in the proof explicit. For example, in order to find a recurrence for $(F_n^2)_{n=0}^{\infty}$, use the Fibonacci recurrence to get the representations

$$F_n^2 = 1F_n^2 + 0F_n F_{n+1} + 0F_{n+1}^2,$$

$$F_{n+1}^2 = 0F_n^2 + 0F_n F_{n+1} + 1F_{n+1}^2,$$

$$F_{n+2}^2 = 1F_n^2 + 2F_n F_{n+1} + 1F_{n+1}^2,$$

$$F_{n+3}^2 = 1F_n^2 + 4F_n F_{n+1} + 4F_{n+1}^2,$$

$$F_{n+4}^2 = 4F_n^2 + 12F_n F_{n+1} + 9F_{n+1}^2.$$

The coefficients c_0, c_1, c_2, c_3, c_4 of the desired recurrence appear in the nullspace vectors of the matrix

$$\begin{pmatrix} 1 & 0 & 1 & 1 & 4 \\ 0 & 0 & 2 & 4 & 12 \\ 0 & 1 & 1 & 4 & 9 \end{pmatrix}.$$

The nullspace here is generated by $(1, -2, -2, 1, 0)$ and $(2, -3, -6, 0, 1)$, corresponding to the two recurrence equations

$$F_n^2 - 2F_{n+1}^2 - F_{n+2}^2 + F_{n+3}^2 = 0 \qquad (n \geq 0)$$

$$\text{and} \quad 2F_n^2 - 3F_{n+1}^2 - 6F_{n+2}^2 + F_{n+4}^2 = 0 \qquad (n \geq 0).$$

Closure properties are extremely useful for systematically proving identities. For example, in order to prove the index duplication formula for Fibonacci numbers,

$$F_{2n} = 2F_n F_{n+1} - F_n^2 \qquad (n \geq 0),$$

all we need to do is to obtain a recurrence for the sequence $(a_n)_{n=0}^{\infty}$ where

$$a_n := F_{2n} - 2F_n F_{n+1} + F_n^2 \qquad (n \geq 0).$$

With the help of Theorem 4.2 (and appropriate computer algebra software), this is easily done. We may find that the sequence $(a_n)_{n=0}^{\infty}$ satisfies the recurrence

$$a_{n+3} = -a_n + 2a_{n+1} + 2a_{n+2} \qquad (n \geq 0),$$

and so in order to show that $a_n = 0$ for all n, it suffices to check that $a_0 = a_1 = a_2 = 0$ and resort to induction on n.

Even better, we do not need to compute the recurrence equations explicitly, but just refer to the bounds on their order that are provided in Theorem 4.2. Given that the Fibonacci numbers F_n satisfy a recurrence of order two, we find without any computation that:

- $(F_{2n})_{n=0}^{\infty}$ satisfies a recurrence of order at most 2,
- $(F_n F_{n+1})_{n=0}^{\infty}$ satisfies a recurrence of order at most 4,
- $(F_n^2)_{n=0}^{\infty}$ satisfies a recurrence of order at most 4,
- Therefore $(2F_n F_{n+1} + F_n^2)_{n=0}^{\infty}$ satisfies a recurrence of order at most 8,
- Therefore $(F_{2n} - 2F_n F_{n+1} - F_n^2)_{n=0}^{\infty}$ satisfies a recurrence of order at most 10.

The index duplication formula for the Fibonacci numbers is therefore proven as soon as it is verified for the finitely many indices $n = 0, 1, \ldots, 9$.

The important feature of this method is that virtually any identity about any C-finite sequences can be reduced mechanically to checking the identity for a finite number of values.

4.4 The Tetrahedron for C-finite Sequences

The most immediate vertex in the Concrete Tetrahedron for C-finite sequences is
the recurrence corner, for the simple reason that C-finite sequences are by definition
those that satisfy a linear recurrence with constant coefficients.

The special form of a C-finite recurrence implies also a special form of the generating function of a C-finite sequence. Starting from a recurrence equation

$$a_{n+r} + c_{r-1}a_{n+r-1} + \cdots + c_0 a_n = 0 \qquad (n \geq 0),$$

multiplying by x^n and summing over all $n \geq 0$ gives

$$\sum_{n=0}^{\infty} a_{n+r}x^n + c_{r-1}\sum_{n=0}^{\infty} a_{n+r-1}x^n + \cdots + c_0 \sum_{n=0}^{\infty} a_n x^n = 0.$$

In terms of the generating function $a(x) = \sum_{n=0}^{\infty} a_n x^n$ of some solution $(a_n)_{n=0}^{\infty}$, this
equation turns into

$$\left(a(x) - (a_0 + a_1 x + \cdots a_{r-1}x^{r-1})\right)/x^r$$
$$+ c_{r-1}\left(a(x) - (a_0 + a_1 x + \cdots + a_{r-2}x^{r-2})\right)/x^{r-1}$$
$$+ \cdots\cdots$$
$$+ c_1\left(a(x) - a_0\right)/x$$
$$+ c_0 a(x) = 0,$$

which is of the form

$$a(x) + c_{r-1}xa(x) + \cdots + c_1 x^{r-1}a(x) + c_0 x^r a(x) = p(x)$$

for some polynomial $p(x)$ of degree at most $r - 1$ that depends on the initial values
$a_0, \ldots, a_{r-1}$ of the sequence $(a_n)_{n=0}^{\infty}$ at hand. We have found that the generating
function $a(x)$ is just a rational function

$$a(x) = \frac{p(x)}{1 + c_{r-1}x + \cdots + c_1 x^{r-1} + c_0 x^r}$$

where the coefficients of the denominator polynomial agree with the coefficients in
the original C-finite recurrence. Going through the steps of the argument in reverse
order shows that every rational function $p(x)/q(x)$ with $\deg p(x) < \deg q(x)$ is the
generating function of some C-finite recurrence:

Theorem 4.3 *A sequence $(a_n)_{n=0}^{\infty}$ in $\mathbb{K}$ satisfies a C-finite recurrence*

$$a_{n+r} + c_{r-1}a_{n+r-1} + \cdots + c_0 a_n = 0 \qquad (n \geq 0)$$

with $c_0, \ldots, c_r \in \mathbb{K}$, $c_0 \neq 0$, if and only if

$$\sum_{n=0}^{\infty} a_n x^n = \frac{p(x)}{1 + c_{r-1}x + \cdots + c_1 x^{r-1} + c_0 x^r}$$

for some polynomial $p(x) \in \mathbb{K}[x]$ of degree at most $r - 1$.

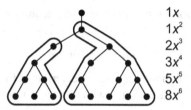

$$
\begin{aligned}
&1x\\
&1x^2\\
&2x^3\\
&3x^4\\
&5x^5\\
&8x^6
\end{aligned}
$$

Fig. 4.2 Self-similarity in the Fibonacci tree

A pictorial interpretation for the generating function $F(x)$ of the Fibonacci numbers is given in Fig. 4.2: as the picture indicates, the family tree of the fertile pairs of rabbits is *self-similar* in the sense that chopping off the root of the tree (corresponding to the linear term x) leaves us with two copies of the family tree, one rooted at level 1 (corresponding to $xF(x)$) and one rooted at level 2 (corresponding to $x^2F(x)$). This motivates the identity

$$F(x) = x + xF(x) + x^2 F(x),$$

from which we get

$$F(x) = \sum_{n=0}^{\infty} F_n x^n = \frac{x}{1 - x - x^2}.$$

Theorem 4.3 also confirms independently some of the closure properties we have proved in the previous section. For instance, the closure of C-finite sequences under addition follows readily from the closure of rational functions under addition and the closure under summation is a direct consequence of the closure of rational functions under multiplication by $1/(1-x)$.

In the case where $\mathbb{K}$ is algebraically closed, also the result about the representation of C-finite sequences in terms of polynomials and exponentials is accessible from the generating function point of view. We know from the previous chapter that the generating function of a polynomial sequence $(p(n))_{n=0}^{\infty}$ is a rational function $q(x)/(1-x)^d$ with $\deg q(x) < d$. The substitution $x \mapsto ux$ for fixed $u \in \mathbb{K}$ reveals that then $q(ux)/(1-ux)^d$ is the generating function of $(p(n)u^n)_{n=0}^{\infty}$. From the fact that every rational function $a(x)/b(x) \in \mathbb{K}(x)$ with $b(0) \neq 0$ and $\deg a(x) < \deg b(x)$ can be brought to the form

$$\frac{a(x)}{b(x)} = \frac{q_1(x)}{(1 - u_1 x)^{d_1}} + \frac{q_2(x)}{(1 - u_2 x)^{d_2}} + \cdots + \frac{q_m(x)}{(1 - u_m x)^{d_m}}$$

for distinct $u_1, \ldots, u_m \in \mathbb{K} \setminus \{0\}$ such that $1/u_i$ is a d_i-th root of $b(x)$ and $q_i(x) \in \mathbb{K}[x]$ is of degree less than d_i (partial fraction decomposition), we obtain an alternative proof of Theorem 4.1.

In the case where $\mathbb{K} \subseteq \mathbb{C}$, the asymptotic behavior of a C-finite sequence can be read off from the partial fraction decomposition of the generating function, or, for that matter, from the closed form representation of the sequence. If we sort exponential terms via

$$n^{d_1} a_1^n \prec n^{d_2} a_2^n \quad :\Longleftrightarrow \quad |a_1| < |a_2| \text{ or } \left(|a_1| = |a_2| \text{ and } d_1 < d_2\right)$$

then the asymptotics of a C-finite sequence is determined by the terms in the closed form representation which are maximal with respect to this order. For example, for $(a_n)_{n=0}^{\infty}$ defined via

$$a_{n+5} - \tfrac{1}{2}a_{n+4} - 2a_{n+3} + a_{n+2} + a_{n+1} - \tfrac{1}{2}a_n = 0 \qquad (n \geq 0),$$
$$\text{and } a_0 = 0, a_1 = 1, a_2 = -1, a_3 = 5, a_4 = 0,$$

we have

$$a_n = -\tfrac{7}{2} - \tfrac{1}{18}(-1)^n + \tfrac{32}{9}(\tfrac{1}{2})^n + \tfrac{7}{4}n - \tfrac{11}{12}n(-1)^n \qquad (n \geq 0),$$

and therefore

$$a_n \sim n\left(\tfrac{7}{4} - \tfrac{11}{12}(-1)^n\right) \qquad (n \to \infty).$$

The exponential terms 1^n and $(-1)^n$ in this asymptotic estimate reflect the fact that 1 and -1 are the poles of the generating function (considered as an analytic function) which are closest to the origin. The polynomial factor n reflects the fact that these poles have multiplicity two.

The closed form representation also gives an immediate access to the summation problem. By the finite version of the geometric series, we have

$$\sum_{k=0}^{n} u^k = \frac{1 - u^{n+1}}{1 - u} \qquad (n \geq 0, u \neq 1),$$

and therefore the sum over an exponential term is essentially again an exponential term. The generating function technique from Sect. 3.3 for summation of polynomial sequences $(p(n))_{n=0}^{\infty}$ admits a straight-forward extension to sequences of the form $(p(n)u^n)_{n=0}^{\infty}$, which provides yet another proof that C-finite sequences are closed under summation.

While this may be acceptable as a theoretical argument, we won't usually proceed along these lines for solving particular summation problems. Most often it will be more interesting to ask whether a given sum over a C-finite sequence can be expressed in terms of the summand sequence itself, like in the identity

$$\sum_{k=0}^{n} F_k = F_n + F_{n+1} - 1 \qquad (n \geq 0).$$

This closed form evaluation is certainly a more comfortable result than some mess of exponentials involving the golden ratio.

Such representations always exist, and it is not hard to write a program that finds them. In order to evaluate a sum $\sum_{k=0}^{n} a_k$ in terms of $a_n, a_{n+1}, \ldots, a_{n+r-1}$, where $(a_n)_{n=0}^{\infty}$ satisfies the recurrence

$$a_{n+r} + c_{r-1}a_{n+r-1} + \cdots + c_1 a_{n+1} + c_0 a_n = 0 \qquad (n \geq 0),$$

all we need to do is to find a linear combination $(b_n)_{n=0}^\infty$ of $(a_n)_{n=0}^\infty, \ldots, (a_{n+r-1})_{n=0}^\infty$ such that $b_{n+1} - b_n = a_n$ for all n, and then apply telescoping. In the example of the Fibonacci identity above, the key step is the observation

$$(F_{n+1} + F_{n+2}) - (F_n + F_{n+1}) = F_{n+1} \qquad (n \geq 0).$$

If the characteristic polynomial $c(x) = x^r + c_{r-1}x^{r-1} + \cdots + c_1 x + c_0$ is such that $c(1) \neq 0$, then we can find by division with remainder a polynomial $q(x) = x^{r-1} + q_{r-2}x^{r-2} + \cdots + q_1 x + q_0 \in \mathbb{K}[x]$ with

$$c(x) = (x-1)q(x) + c(1).$$

Setting

$$(b_n)_{n=0}^\infty := q(x) \bullet (a_n)_{n=0}^\infty = (a_{n+r-1} + q_{r-2}a_{n+r-2} + \cdots + q_1 a_{n+1} + q_0 a_n)_{n=0}^\infty,$$

the equation $c(x) = (x-1)q(x) + c(1)$ translates into

$$b_{n+1} - b_n + c(1)a_n = 0 \qquad (n \geq 0),$$

and telescoping gives

$$\sum_{k=0}^n a_k = -\frac{1}{c(1)}(b_{n+1} - b_0) \qquad (n \geq 0),$$

which is of the desired form.

If $c(1) = 0$, we can still write $c(x) = (x-1)^m \bar{c}(x)$ with $m \geq 1$ and $\bar{c}(x) \in \mathbb{K}[x]$ such that $\bar{c}(1) \neq 0$, and we will find some $\bar{q}(x) = x^{r-m-1} + \bar{q}_{r-m-2}x^{r-m-2} + \cdots + \bar{q}_0 \in \mathbb{K}[x]$ with

$$c(x) = (x-1)^m \big((x-1)\bar{q}(x) + \bar{c}(1)\big).$$

Defining $(b_n)_{n=0}^\infty := \bar{q}(x) \bullet (a_n)_{n=0}^\infty$, this equation translates into

$$\Delta^m \big(b_{n+1} - b_n + \bar{c}(1)a_n\big) = 0 \qquad (n \geq 0),$$

where Δ is the forward difference operator. These m difference operators can be undone by m repeated summations. After a first summation, we find that the sequence $\Delta^{m-1}(b_{n+1} - b_n + \bar{c}(1)a_n)_{n=0}^\infty$ is constant. Summing this constant sequence once more, we find that the sequence $\Delta^{m-2}(b_{n+1} - b_n + \bar{c}(1)a_n)_{n=0}^\infty$ must be a polynomial sequence of degree one, and so on. After m summations, we eventually find that

$$b_{n+1} - b_n + \bar{c}(1)a_n = \bar{p}(n) \qquad (n \geq 0)$$

for some polynomial $\bar{p}(x) \in \mathbb{K}[x]$ of degree at most $m-1$. One final summation then gives the representation

$$\sum_{k=0}^n a_k = -\frac{1}{\bar{c}(1)}b_{n+1} + p(n) \qquad (n \geq 0)$$

for some polynomial $p(x) \in \mathbb{K}[x]$ of degree at most m.

4.5 Systems of C-finite Recurrences

A C-finite system of recurrence equations is a simultaneous equation for several sequences $(a_n^{(1)})_{n=0}^{\infty}, \ldots, (a_n^{(s)})_{n=0}^{\infty}$ of the form

$$
\begin{pmatrix} a_{n+r}^{(1)} \\ a_{n+r}^{(2)} \\ \vdots \\ a_{n+r}^{(s)} \end{pmatrix} + C_{r-1} \begin{pmatrix} a_{n+r-1}^{(1)} \\ a_{n+r-1}^{(2)} \\ \vdots \\ a_{n+r-1}^{(s)} \end{pmatrix} + \cdots + C_1 \begin{pmatrix} a_{n+1}^{(1)} \\ a_{n+1}^{(2)} \\ \vdots \\ a_{n+1}^{(s)} \end{pmatrix} + C_0 \begin{pmatrix} a_n^{(1)} \\ a_n^{(2)} \\ \vdots \\ a_n^{(s)} \end{pmatrix} = 0 \quad (n \geq 0)
$$

where $C_0, \ldots, C_{r-1} \in \mathbb{K}^{s \times s}$ are matrices and C_0 is invertible. Clearly, each C-finite sequence is also a solution of such a system, as the special case $s = 1$ reduces to the usual C-finite recurrences. But for $s > 0$, the system may relate the values of the s sequences mutually to each other. For example, the system

$$
\begin{aligned}
u_{n+1} &= u_n + 3v_n \quad (n \geq 0), \\
v_{n+1} &= 2u_n + 2v_n \quad (n \geq 0),
\end{aligned}
$$

is of the form we are considering. (For clarity, the matrix-vector multiplication is explicitly spelled out here.) It is natural to ask whether this additional freedom allows for defining sequences that are not C-finite in the usual sense, that is, whether the solutions of C-finite systems form a larger class of sequences than the solutions of (scalar) C-finite recurrences do.

It turns out that this is not the case. To see this, observe first that we can restrict ourselves without loss of generality to first order systems, because a system of order r is equivalent to the first order system

$$
\begin{pmatrix} A_{n+1} \\ A_{n+2} \\ \vdots \\ \vdots \\ A_{n+r} \end{pmatrix} + \begin{pmatrix} 0 & I_s & 0 & \cdots & 0 \\ \vdots & 0 & \ddots & \ddots & \vdots \\ \vdots & \vdots & \ddots & \ddots & 0 \\ 0 & 0 & \cdots & 0 & I_s \\ C_0 & C_1 & \cdots & \cdots & C_{r-1} \end{pmatrix} \begin{pmatrix} A_n \\ A_{n+1} \\ \vdots \\ \vdots \\ A_{n+r-1} \end{pmatrix} = 0 \quad (n \geq 0),
$$

where I_s refers to the $s \times s$ unit matrix and A_{n+i} refers to the vector $(a_{n+i}^{(1)}, \ldots, a_{n+i}^{(s)})$. So let us consider a first order system

$$
A_{n+1} = MA_n \quad (n \geq 0),
$$

where M is some invertible $s \times s$ matrix over $\mathbb{K}$. If we unfold this recurrence like a scalar recurrence, we get

$$
A_n = M^n A_0 \quad (n \geq 0),
$$

where M^n is the n-fold matrix multiplication of M with itself.

By the theorem of Cayley-Hamilton, the characteristic polynomial $c(x) := \det(M - xI_s)$ of the matrix M has the property that $c(M) = 0$, therefore, if $c(x) = c_0 + c_1 x + \cdots + c_s x^s$, then

$$c_s M^s + c_{s-1} M^{s-1} + \cdots + c_1 M + c_0 I_s = 0.$$

Multiplying this from the left to $M^n A_0$, we find

$$c_s A_{n+s} + c_{s-1} A_{n+s-1} + \cdots + c_1 A_{n+1} + c_0 A_n = 0 \quad (n \geq 0).$$

This means that all the coordinate sequences $(a_n^{(i)})_{n=0}^\infty$ satisfy the same recurrence equation

$$c_s a_{n+s}^{(i)} + c_{s-1} a_{n+s-1}^{(i)} + \cdots + c_1 a_{n+1}^{(i)} + c_0 a_n^{(i)} = 0 \quad (n \geq 0),$$

so in particular they are C-finite.

Let us use this insight to determine the solutions of the example system

$$\begin{pmatrix} u_{n+1} \\ v_{n+1} \end{pmatrix} = \begin{pmatrix} 1 & 3 \\ 2 & 2 \end{pmatrix} \begin{pmatrix} u_n \\ v_n \end{pmatrix} \quad (n \geq 0).$$

The characteristic polynomial of the matrix M in this example is $(x-4)(x+1)$, and therefore any solution (u_n, v_n) will have the form

$$\begin{pmatrix} u_n \\ v_n \end{pmatrix} = \begin{pmatrix} c_{1,1} \\ c_{2,1} \end{pmatrix} 4^n + \begin{pmatrix} c_{1,2} \\ c_{2,2} \end{pmatrix} (-1)^n \quad (n \geq 0)$$

for certain constants $c_{1,1}, c_{1,2}, c_{2,1}, c_{2,2}$. But not every choice of constants gives rise to solutions of the system. To find out which ones do, we can simply substitute the general form of the solution into the original system and compare the coefficients of 4^n and $(-1)^n$ to zero. This gives constraints on the $c_{i,j}$. The requirement

$$\begin{pmatrix} c_{1,1} \\ c_{2,1} \end{pmatrix} 4^{n+1} + \begin{pmatrix} c_{1,2} \\ c_{2,2} \end{pmatrix} (-1)^{n+1} \stackrel{!}{=} \begin{pmatrix} 1 & 3 \\ 2 & 2 \end{pmatrix} \left(\begin{pmatrix} c_{1,1} \\ c_{2,1} \end{pmatrix} 4^n + \begin{pmatrix} c_{1,2} \\ c_{2,2} \end{pmatrix} (-1)^n \right)$$

implies the condition

$$\begin{pmatrix} 3c_{1,1} - 3c_{2,1} \\ -2c_{1,1} + 2c_{2,1} \end{pmatrix} 4^n + \begin{pmatrix} -2c_{1,2} - 3c_{2,2} \\ -2c_{1,2} - 3c_{2,2} \end{pmatrix} (-1)^n \stackrel{!}{=} 0.$$

Forcing the coefficient vectors to zero gives the linear system

$$\begin{pmatrix} 3 & 0 & -3 & 0 \\ -2 & 0 & 2 & 0 \\ 0 & -2 & 0 & -3 \\ 0 & -2 & 0 & -3 \end{pmatrix} \begin{pmatrix} c_{1,1} \\ c_{1,2} \\ c_{2,1} \\ c_{2,2} \end{pmatrix} = 0$$

whose solution space is generated by the vectors $(1,0,1,0)$ and $(0,-3,0,2)$. Consequently, the solution space of the original C-finite recurrence system consists precisely of the $\mathbb{K}$-linear combinations of

$$\begin{pmatrix} 1 \\ 1 \end{pmatrix} 4^n \quad \text{and} \quad \begin{pmatrix} -3 \\ 2 \end{pmatrix} (-1)^n.$$

The vectors $(1,1)$ and $(-3,2)$ appearing here are eigenvectors of M for the eigenvalues 4 and -1, respectively. Of course, this is not a coincidence but a rule.

Before concluding this section, we still want to mention that C-finite systems can also be approached by generating functions. For $i = 1,\ldots,s$, let $a^{(i)}(x) := \sum_{n=0}^{\infty} a_n^{(i)} x^n$ be the generating functions of the sequences arising in a system $A_{n+1} = MA_n$ as above. The various recurrence equations

$$a_{n+1}^{(i)} = m_{i,1} a_n^{(1)} + m_{i,2} a_n^{(2)} + \cdots + m_{i,s} a_n^{(s)} \qquad (n \geq 0, i = 1,\ldots,s)$$

that make up this system correspond to equations

$$a^{(i)}(x) - a^{(i)}(0) = x m_{i,1} a^{(1)}(x) + x m_{i,2} a^{(2)}(x) + \cdots + x m_{i,s} a^{(s)}(x) \qquad (i = 1,\ldots,s)$$

for the generating functions $a^{(1)}(x),\ldots,a^{(s)}(x)$. These equations form a linear system of equations over $\mathbb{K}(x)$ which has a unique solution of the form

$$\begin{pmatrix} a^{(1)}(x) \\ \vdots \\ a^{(s)}(x) \end{pmatrix} = \frac{1}{c_s + c_{s+1}x + \cdots + c_0 x^s} \begin{pmatrix} q^{(1)}(x) \\ \vdots \\ q^{(s)}(x) \end{pmatrix}$$

where $c_0 + c_1 x + \cdots + c_s x^s = \det(M - xI_s) \in \mathbb{K}[x]$ is the characteristic polynomial and $q^{(1)}(x),\ldots,q^{(s)}(x) \in \mathbb{K}[x]$ are some polynomials of degree less than s.

4.6 Applications

Regular Languages

In the theory of formal languages, a *language* is a set of *words* consisting of some *letters* chosen from a fixed *alphabet*. Words are simply tuples of letters, for example, if the alphabet is $\Sigma = \{a,b,c\}$, then the language Σ^* of all the words is

$$\Sigma^* = \{\varepsilon, a, b, c, aa, ab, ac, ba, bb, bc, ca, cb, cc, aaa, aab, aac, \ldots\},$$

where ε refers to the empty word, a word that has no letters at all. An example for a nontrivial language is the language $L \subseteq \Sigma^*$ of all words that do not contain aa or bb or cc as subwords:

$$L = \{\varepsilon, a, b, c, ab, ac, ba, bc, ca, cb, aba, abc, aca, acb, bab, \ldots\} \subseteq \Sigma^*.$$

Language theorists have come up with a hierarchy of languages that allows to rank formal languages according to how difficult they are. One of the most elementary types of languages is the class of regular languages. These are languages that can be constructed inductively, by saying that:

- The empty language $\varnothing$ and the languages $\{\omega\}$ containing a single word $\omega \in \Sigma^*$ are regular languages.

- If A and B are regular languages, then so are their union $A \cup B$ as well as

$$A.B := \{ab : a \in A, b \in B\}$$
$$\text{and} \quad A^* := \{\varepsilon\} \cup A \cup A.A \cup A.A.A \cup A.A.A.A \cup \cdots,$$

where ab refers to the word obtained from a and b by joining them together.

Regular languages are used in computer science for describing the most elementary syntactic entities of a programming language. For example, if floating point numbers are written according to the usual rules as words over the alphabet $\{+, -, 0, 1, 2, 3, 4, 5, 6, 7, 8, 9, ., e\}$, then the set of words that are syntactically correct encodings of floating point numbers forms a regular language. It is less obvious, but also true, that the language L defined before is also a regular language.

The relation of regular languages to C-finite sequences is that the number of words with exactly n letters belonging to a given regular language is always a C-finite sequence. The language L, for example, contains exactly $3 \cdot 2^n$ words of length $n \geq 1$.

Unrestricted Lattice Walks

Multivariate recurrence equations with constant coefficients arise for instance in enumeration problems related to lattice walks. As an example, consider walks in $\mathbb{Z}^2$ starting in the origin $(0,0)$ and consisting of n steps where each single step can go either north-east ($\nearrow$) or south ($\downarrow$) or west ($\leftarrow$). One such walk with $n = 10$ steps is shown in Fig. 4.3.

We are interested in the number $a_{n,i,j}$ of walks with n steps that end at a given point $(i, j) \in \mathbb{Z}^2$. Obviously, we can compute this number recursively via

$$a_{n+1,i,j} = a_{n,i-1,j-1} + a_{n,i+1,j} + a_{n,i,j+1} \qquad (n \geq 0; i, j \in \mathbb{Z}),$$

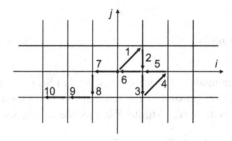

Fig. 4.3 A walk in the plane consisting of ten steps

the initial conditions being given by

$$a_{0,i,j} = \begin{cases} 1 & \text{if } i = j = 0 \\ 0 & \text{otherwise} \end{cases}.$$

It is clear that by making n steps we cannot reach any point (i,j) with $|i| > n$ or $|j| > n$, so we have $a_{n,i,j} = 0$ for these i,j. Therefore, for every fixed n, the doubly infinite series

$$\sum_{i=-\infty}^{\infty} \sum_{j=-\infty}^{\infty} a_{n,i,j} x^i y^j \qquad (n \in \mathbb{N} \text{ fixed})$$

is actually just a rational function in x and y. The generating function

$$a(t,x,y) := \sum_{n=0}^{\infty} \left(\sum_{i \in \mathbb{Z}} \sum_{j \in \mathbb{Z}} a_{n,i,j} x^i y^j \right) t^n$$

can therefore be considered as an element of $\mathbb{Q}(x,y)[[t]]$.

A straight-forward calculation starting from the multivariate recurrence equation for the $a_{n,i,j}$ leads to the equation

$$\frac{1}{t} \left(a(t,x,y) - 1 \right) = xya(t,x,y) + \frac{1}{x}a(t,x,y) + \frac{1}{y}a(t,x,y)$$

for the generating function, from which we directly obtain

$$a(t,x,y) = \frac{1}{1 - t(xy + \frac{1}{x} + \frac{1}{y})} = \sum_{n=0}^{\infty} \left(xy + \frac{1}{x} + \frac{1}{y} \right)^n t^n$$

$$= 1 + (xy + x^{-1} + y^{-1})t + (x^{-2} + 2x + y^{-2} + 2x^{-1}y^{-1} + 2y + x^2y^2)t^2 + \cdots$$

For example, the coefficient 2 in front of the term xt^2 ($= x^1y^0t^2$) means that there are precisely two ways to get in two steps from the origin to the point $(1,0)$. Observe also that setting $x = y = 1$ (which in this special situation is allowed) gives the geometric series $1/(1 - 3t)$, in accordance with the fact that there are in total 3^n walks with n steps and arbitrary endpoint.

Chebyshev Polynomials

The Chebyshev polynomials $T_n(x)$ were introduced in Sect. 3.5 as the polynomials appearing in the series expansion of the rational function $(1 - xy)/(1 - 2xy + y^2)$ with respect to y, hence they are C-finite. We have the recurrence relation

$$T_{n+2}(x) - 2xT_{n+1}(x) + T_n(x) = 0 \qquad (n \geq 0).$$

Chebyshev polynomials appear in the repeated application of the trigonometric addition theorem for cosine: We have $\cos(nt) = T_n(\cos t)$ $(n \geq 0)$, because this is true for $n = 0$ and $n = 1$, and the recurrence holds because of

$$\cos((n+1)t+t) + \cos((n+1)t-t) = 2\cos(t)\cos((n+1)t) = 0 \qquad (n \geq 0)$$

by applying the addition theorem for cosine twice.

The substitution $x = \cos t$ also allows for a quick proof of the orthogonality relation of Chebyshev polynomials (Sect. 3.5): Applying the substituting $x = \cos t$, $dx = -\sin t \, dt$ to the integral

$$\int_{-1}^{1} \frac{T_m(x) T_n(x)}{\sqrt{1-x^2}} dx$$

gives

$$\int_{\pi}^{0} \frac{T_m(\cos t) T_n(\cos t)}{\sin t} (-\sin t) dt = \int_{0}^{\pi} \cos(mt) \cos(nt) \, dt.$$

For this integral, the desired result is easily obtained by repeated integration by parts.

Chebyshev polynomials are relevant in numerical analysis. Here, instead of expanding an analytic function $f \colon \mathbb{C} \to \mathbb{C}$ in a Taylor series about the origin, one considers expansions of f as a series in terms of Chebyshev polynomials:

$$f(z) = \sum_{n=0}^{\infty} a_n T_n(z).$$

The coefficients in this expansion are easily determined thanks to the orthogonality relation:

$$\int_{-1}^{1} \frac{f(z) T_n(z)}{\sqrt{1-z^2}} dz = \int_{-1}^{1} \sum_{k=0}^{\infty} a_k T_k(z) T_n(z) \frac{dz}{\sqrt{1-z^2}}$$

$$= \sum_{k=0}^{\infty} a_k \int_{-1}^{1} T_k(z) T_n(z) \frac{dz}{\sqrt{1-z^2}} = a_n \pi$$

for $n \geq 1$. (For $n = 0$, the integral exceptionally evaluates to $a_n \pi / 2$.) This works analogously for any family of orthogonal polynomials.

The numerical interest in Chebyshev expansions is rooted in the feature that the polynomial $p_n(x) = \sum_{k=0}^{n} a_k T_k(x)$ obtained by truncating the series after the n-th term is such that

$$\max_{z \in [-1,1]} |f(z) - p_n(z)|$$

becomes almost as small as it can possibly get for a polynomial of degree n. See [48] for precise estimation statements and proofs.

For example, a pretty good approximation of the exponential function is $p_2(x) = a_0 + a_1 T_1(x) + a_2 T_2(x)$, where

$$a_0 = \frac{2}{\pi} \int_{-1}^{1} \frac{\exp(z) T_0(z)}{\sqrt{1-z^2}} dz \approx 1.26607,$$

$$a_1 = \frac{1}{\pi} \int_{-1}^{1} \frac{\exp(z) T_1(z)}{\sqrt{1-z^2}} dz \approx 1.13032,$$

$$a_2 = \frac{1}{\pi} \int_{-1}^{1} \frac{\exp(z) T_2(z)}{\sqrt{1-z^2}} dz \approx 0.271495.$$

The polynomial $p_2(z)$ is plotted together with $\exp(z)$ in Fig. 4.4. For the next polynomial $p_3(z)$, a difference to the curve of $\exp(z)$ would already be hardly visible.

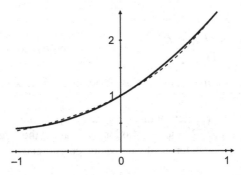

Fig. 4.4 $\exp(z)$ (dashed) and the quadratic polynomial (solid) that comes closest to it on $[-1, 1]$

4.7 Problems

Problem 4.1 Determine the first ten decimal digits of $F_{2^{1000}}$.

Problem 4.2 Determine the last ten decimal digits of $F_{2^{1000}}$.

Problem 4.3 Express $\sum_{k=0}^{n} F_{n+k}$ in terms of Fibonacci numbers.

Problem 4.4 Prove the identity $\sum_{k=0}^{n} \binom{n-k}{k} = F_{n+1}$ $(n \geq 0)$.

(*Hint:* Recall that $1/(1-x-xy) = \sum_{n,k=0}^{\infty} \binom{n}{k} x^n y^k$ and consider the substitution $y \mapsto x$ in the sense of power series composition.)

Problem 4.5 Prove parts 4 and 5 of Theorem 4.2.

Problem 4.6 Consider the continued fraction

$$1 + \cfrac{1}{1 + \cfrac{1}{1 + \cfrac{1}{1 + \cdots}}}.$$

What is the value of the finite continued fraction obtained by truncating at the n-th level? What is the value of the infinite continued fraction?

Problem 4.7 (Vandermonde's determinant) For $a_1, \ldots, a_m \in \mathbb{K}$, consider

$$A := \begin{pmatrix} 1 & 1 & \cdots & 1 \\ a_1 & a_2 & \cdots & a_m \\ \vdots & \vdots & & \vdots \\ a_1^{m-1} & a_2^{m-1} & \cdots & a_m^{m-1} \end{pmatrix} \in \mathbb{K}^{m \times m}.$$

Use the determinant formula

$$\det A = \prod_{1 \le i < j \le m} (a_i - a_j)$$

to show:

1. For any pairwise different numbers $a_i \in \mathbb{K}$ ($i = 1, \ldots, m$), the sequences $(a_i^n)_{n=0}^{\infty}$ are linearly independent over $\mathbb{K}$.

2. For any pairwise different numbers $a_i \in \mathbb{K}$ ($i = 1, \ldots, m$) and any numbers $b_i \in \mathbb{K}$ ($i = 1, \ldots, m$), there exists precisely one polynomial $p(x) \in \mathbb{K}[x]$ of degree $m - 1$ (or less) such that $p(a_i) = b_i$ ($i = 1, \ldots, m$).

Problem 4.8 Let $(a_n)_{n=0}^{\infty}$ be a C-finite sequence and suppose that $(u_n)_{n=0}^{\infty}$ satisfies the inhomogeneous linear recurrence

$$u_{n+r} + c_{r-1} u_{n+r-1} + \cdots + c_1 u_{n+1} + c_0 u_n = a_n \qquad (n \ge 0).$$

Prove that $(u_n)_{n=0}^{\infty}$ is C-finite.

Problem 4.9 Show that $(H_n)_{n=0}^{\infty}$ is not C-finite.

Problem 4.10 Show that $(S_2(n,k))_{n=0}^{\infty}$ is C-finite for every fixed $k \in \mathbb{N}$.

Problem 4.11 The goal of this problem is to show that for a polynomial $p(x,y) \in \mathbb{Q}[x,y]$ we have $p(F_n, F_{n+1}) = 0$ for all $n \geq 0$ if and only if $p(x,y)$ is a multiple of $u(x,y)^2 - 1$ where $u(x,y) = x^2 - xy - y^2$.

1. Show that $u(F_{2n}, F_{2n+1}) = 1$ and $u(F_{2n+1}, F_{2n}) = -1$ for all $n \in \mathbb{N}$. Conclude the implication from right to left.

2. Show that for every polynomial $p(x,y)$ there is a polynomial $a(x,y)$ such that $q(x,y) := p(x,y) + (u(x,y) - 1)a(x,y)$ is linear in y. Conclude that $p(F_{2n}, F_{2n+1}) = 0$ for all $n \in \mathbb{N}$ if and only if $q(F_{2n}, F_{2n+1}) = 0$ for all $n \in \mathbb{N}$.

3. Show by an asymptotic argument that if two polynomials $q_0(x), q_1(x) \in \mathbb{Q}[x]$ are such that $q_0(F_{2n}) + q_1(F_{2n})F_{2n+1} = 0$ for all $n \in \mathbb{N}$ then $q_0(x) = q_1(x) = 0$. Conclude that $p(F_{2n}, F_{2n+1}) = 0$ for all $n \in \mathbb{N}$ if and only if $p(x,y)$ is a multiple of $u(x,y) - 1$.

4. Show that $p(F_{2n+1}, F_{2n}) = 0$ for all $n \in \mathbb{N}$ if and only if $p(x,y)$ is a multiple of $u(x,y) + 1$ by an analogous argument. Conclude that $p(F_n, F_{n+1}) = 0$ for all $n \in \mathbb{N}$ if and only if $p(x,y)$ is a multiple of $u(x,y)^2 - 1$.

Problem 4.12 We have observed that the solution $(a_n)_{n=0}^\infty$ of a C-finite recurrence of order r is uniquely determined by the initial values $a_0, \ldots, a_{r-1}$. More generally, let $i_1, \ldots, i_r \in \mathbb{N}$ be pairwise distinct. Is the solution $(a_n)_{n=0}^\infty$ uniquely determined by the r values $a_{i_1}, \ldots, a_{i_r}$?

Problem 4.13 The *exponential generating function* $\bar{a}(x)$ of a sequence $(a_n)_{n=0}^\infty$ is defined as the formal power series $\sum_{n=0}^\infty a_n \frac{x^n}{n!}$. Show that a sequence satisfies a linear recurrence with constant coefficients if and only if its exponential generating function satisfies a linear differential equation with constant coefficients.

Problem 4.14 Suppose that the characteristic polynomial of the C-finite recurrence

$$a_{n+r} - c_{r-1}a_{n+r-1} - \cdots - c_0 a_n = 0 \qquad (n \geq 0)$$

has r pairwise different roots $u_1, \ldots, u_r \in \mathbb{K} \setminus \{0\}$, and let (like in Sect. 4.5 with $s = 1$)

$$M = \begin{pmatrix} 0 & 1 & 0 & \cdots & 0 \\ \vdots & 0 & \ddots & \ddots & \vdots \\ \vdots & \vdots & \ddots & \ddots & 0 \\ 0 & 0 & \cdots & 0 & 1 \\ c_0 & c_1 & \cdots & \cdots & c_{r-1} \end{pmatrix}.$$

Show that

$$M = \begin{pmatrix} 1 & \cdots & 1 \\ u_1 & \cdots & u_r \\ \vdots & & \vdots \\ u_1^{r-1} & \cdots & u_r^{r-1} \end{pmatrix} \begin{pmatrix} u_1 & 0 & \cdots & 0 \\ 0 & u_2 & \ddots & \vdots \\ \vdots & & \ddots & 0 \\ 0 & \cdots & 0 & u_r \end{pmatrix} \begin{pmatrix} 1 & \cdots & 1 \\ u_1 & \cdots & u_r \\ \vdots & & \vdots \\ u_1^{r-1} & \cdots & u_r^{r-1} \end{pmatrix}^{-1}$$

and use this result to obtain an alternative proof of Theorem 4.1 for the special case under consideration.

Chapter 5
Hypergeometric Series

Hypergeometric series form a particularly powerful class of series, as they appear in a great variety of different scientific contexts while at the same time allowing a rather simple definition. Popularized by the work of Gauss, hypergeometric series have been intensively studied since the 19th century, and they are still subject of ongoing research. Nowadays, they are also well understood from an algorithmic point of view, and in this chapter, we will see some of the most important algorithms for dealing with them.

5.1 The Binomial Theorem

The binomial theorem states that for $n \in \mathbb{N}$ and any $a, b \in \mathbb{K}$ we have

$$(a+b)^n = \sum_{k=0}^{n} \binom{n}{k} a^k b^{n-k}.$$

If this is to be a theorem, there should be a proof. In order to give a rigorous proof, we need to go back to the definition of the symbol $\binom{n}{k}$. Where the binomial coefficients appeared in earlier chapters of this book, we have silently considered them as known and did not give a formal definition. Now it is time to close this gap.

There are several possibilities. We could simply take the binomial theorem itself as the definition. We could as well declare that $\binom{n}{k}$ be defined recursively via the Pascal triangle recurrence

$$\binom{n}{k} = \binom{n-1}{k} + \binom{n-1}{k-1} \quad (n, k > 0)$$

together with the boundary conditions $\binom{n}{0} = 1$ $(n \geq 0)$ and $\binom{0}{n} = 0$ $(n > 0)$. We could instead also take a combinatorial approach and decide that $\binom{n}{k}$ should be the number of subsets of $\{1, 2, \ldots, n\}$ with exactly k elements (hence the pronunciation

M. Kauers, P. Paule, *The Concrete Tetrahedron*
© Springer-Verlag/Wien 2011

"n choose k" for $\binom{n}{k}$). But all these definitions are somehow limited to the case of n and k being natural numbers.

As it is sometimes useful to have binomial coefficients defined for other domains as well, we will adopt right from the outset a definition that covers most of the cases that we will encounter: Let R be some commutative ring containing $\mathbb{Q}$ and let $\lambda \in R$ and $k \in \mathbb{Z}$. Then

$$\binom{\lambda}{k} := \frac{\lambda^{\underline{k}}}{k!} = \frac{\lambda(\lambda-1)\cdots(\lambda-k+1)}{k(k-1)\cdots 1} \qquad (k \geq 0)$$

and $\binom{\lambda}{k} := 0$ for $k < 0$.

When the upper argument is an integer, this general definition implies all the previously stated properties, including the binomial theorem (Problem 5.1). Written in the form

$$(1+x)^{\lambda} = \sum_{n=0}^{\infty} \binom{\lambda}{n} x^n,$$

the binomial theorem continues to hold even in cases when λ is not a natural number. For example, for $\lambda = -1$ we obtain

$$\frac{1}{1+x} = \sum_{n=0}^{\infty} \binom{-1}{n} x^n = \sum_{n=0}^{\infty} (-1)^n x^n,$$

which is correct as an identity in $\mathbb{Q}[[x]]$. It can be checked that this is so for any negative integer in place of λ. For $\lambda = \frac{1}{2}$, we obtain the identity

$$\sqrt{1+x} = \sum_{n=0}^{\infty} \binom{1/2}{n} x^n = \sum_{n=0}^{\infty} \frac{(1/2)^{\underline{n}}}{n!} x^n,$$

which is correct as an identity in $\mathbb{Q}[[x]]$ in the sense that the series on the right satisfies the defining equation of the square root,

$$\left(\sum_{n=0}^{\infty} \frac{(1/2)^{\underline{n}}}{n!} x^n \right)^2 = 1+x.$$

Again, this generalizes and it can be checked that this works as well for any rational number in place of λ.

If $\lambda \in \mathbb{C}$ is an irrational number, then $(1+x)^{\lambda}$ does a priori not have a meaning as a formal power series, but we still have that

$$(1+z)^{\lambda} = \sum_{n=0}^{\infty} \binom{\lambda}{n} z^n \qquad (z \in \mathbb{C}, |z| < 1)$$

as analytic power series. For these λ, and more generally when λ is any element from a commutative ring R, it is therefore reasonable to take

$$(1+x)^{\lambda} := \sum_{n=0}^{\infty} \binom{\lambda}{n} x^n \in R[[x]]$$

as the definition of the symbol $(1+x)^\lambda$. With this general definition, we can prove, for instance, the multiplication law

$$(1+x)^\lambda (1+x)^\mu = (1+x)^{\lambda+\mu}$$

in $R[[x]]$. For, applying the definition and comparing coefficients, this identity is equivalent to the summation formula

$$\sum_{k=0}^{n} \binom{\lambda}{k}\binom{\mu}{n-k} = \binom{\lambda+\mu}{n} \qquad (n \geq 0, \lambda, \mu \in R),$$

which is known as Vandermonde convolution. To see that this identity is valid, assume temporarily that $\lambda = u$ and $\mu = v$, where u, v are some natural numbers. Consider an urn with $u + v$ distinct balls, say u green balls and v blue balls. Then both sides of the identity count the number of ways to choose n balls from the urn, so the identity is evident. Now view λ and μ as indeterminates. Then both sides of the identity are polynomials in λ and μ of total degree n. Since these polynomials evaluate to the same number at every point $(u, v) \in \mathbb{N}^2$ (by the combinatorial argument), they must be identical (Problem 5.2). The identity being true for λ, μ regarded as indeterminates, it will remain true whatever we may substitute for λ and μ, so the identity is true in general.

Of course, there are various other ways to prove the Vandermonde identity, and we will see in Sect. 5.5 that there is even a systematic algorithm for proving summation identities about binomial coefficients so that we can actually have such identities proven by a computer algebra system. Before the invention of this algorithm, proving summation identities about binomial coefficients was widely feared as a hazardous and laborious enterprise and depended on firm knowledge of a large repertoire of basic formulas such as

$$\binom{\lambda}{k} = \binom{\lambda-1}{k} + \binom{\lambda-1}{k-1} \quad (k \in \mathbb{Z}, \lambda \in R) \qquad \text{(Pascal triangle)},$$

$$\binom{\lambda}{k} = (-1)^k \binom{k-\lambda-1}{k} \quad (k \in \mathbb{Z}, \lambda \in R) \qquad \text{(reflection formula)},$$

$$\binom{\lambda}{k} k = \lambda \binom{\lambda-1}{k-1} \quad (k \in \mathbb{Z}, \lambda \in R) \qquad \text{(absorption)}$$

$$\binom{\lambda}{k+1} = \frac{\lambda-k}{k+1}\binom{\lambda}{k} \quad (k \in \mathbb{Z} \setminus \{-1\}, \lambda \in R) \qquad \text{(shift in } k),$$

$$\binom{\lambda+1}{k} = \frac{\lambda+1}{\lambda-k+1}\binom{\lambda}{k} \quad (k \in \mathbb{Z} \setminus \{\lambda+1\}, \lambda \in R) \qquad \text{(shift in } \lambda),$$

and many many others. There were wizards who could skillfully use these formulas to prove virtually any binomial summation identity presented to them by a direct calculation. (Chap. 5 of Concrete Mathematics has a great tutorial on how to deal with binomial coefficients "by hand".) As a more standardized approach to hyper-geometric summation, Andrews, Askey and their hypergeometric school promoted the systematic use of hypergeometric series transformations, which we briefly touch in Sect. 5.3 below. This method reliably guides experts to simple proofs of compli-cated identities, but it also depends on experience and cannot be easily implemented on a computer. Fully algorithmic approaches to summation are also available, and will be discussed in Sects. 5.4 and 5.5. Although these algorithms render hand-calculations obsolete in many situations, it is still useful to gain some practice in manipulating binomial coefficient expressions, as they appear often enough in com-bination with other quantities for which no automatic proof machinery is available yet.

5.2 Basic Facts and Definitions

The binomial coefficients $\binom{n}{k}$ form a bivariate hypergeometric sequence. But let us start with the univariate case. A univariate sequence $(a_n)_{n=0}^\infty$ in $\mathbb{K}$ is called *hyperge-ometric* if there exists a rational function $r(x) \in \mathbb{K}(x)$ such that

$$a_{n+1} = r(n)a_n$$

for all $n \in \mathbb{N}$ where $r(n)$ is defined. We then call $r(x)$ the *shift quotient* of $(a_n)_{n=0}^\infty$. If $r(n)$ is defined for all $n \in \mathbb{N}$, then unwinding the recurrence down to the initial value a_0 yields the form

$$a_n = r(n-1)r(n-2)r(n-3)\cdots r(0)a_0 \quad (n \in \mathbb{N}).$$

For example, every geometric sequence $(a^n)_{n=0}^\infty$ ($a \in \mathbb{K}$ fixed) is hypergeometric (take $r(x) = a$), the sequence $(n)_{n=0}^\infty$ is hypergeometric (take $r(x) = \frac{x+1}{x}$), and the sequence $(n!)_{n=0}^\infty$ is hypergeometric (take $r(x) = x+1$). More generally, for every fixed $z \in \mathbb{N}$, the sequence $((z+n)!)_{n=0}^\infty$ is hypergeometric (take $r(x) = x+z+1$), and, even more generally, for every fixed $z_1, \ldots, z_p \in \mathbb{N}$ and $w_1, \ldots, w_q \in \mathbb{N}$ and $c \in \mathbb{K}$, the sequence $(a_n)_{n=0}^\infty$ with

$$a_n = c^n \frac{(z_1+n)!(z_2+n)!\cdots(z_p+n)!}{(w_1+n)!(w_2+n)!\cdots(w_q+n)!} \quad (n \geq 0)$$

is hypergeometric because

$$a_{n+1} = c\frac{(n+z_1+1)(n+z_2+1)\cdots(n+z_p+1)}{(n+w_1+1)(n+w_2+1)\cdots(n+w_q+1)}a_n \quad (n \geq 0).$$

Taking into account that over an algebraically closed field $\mathbb{K}$ we can always break the numerator and denominator of a rational into linear factors, this is already pretty close to the most general situation. Just that for an arbitrary rational function the z_i and w_j will in general not be natural numbers. Where they are not, we cannot write the hypergeometric sequence in terms of factorials before we agree what, say, $(\frac{1}{2})!$ is supposed to be. One possibility is to use rising (or falling) factorials. From $a^{\overline{n+1}}/a^{\overline{n}} = a + n$ we get the representation

$$a_n = c^n \frac{z_1^{\overline{n}} z_2^{\overline{n}} \cdots z_p^{\overline{n}}}{w_1^{\overline{n}} w_2^{\overline{n}} \cdots w_q^{\overline{n}}}$$

which, however, is also not always defined either. For instance, since $(-1)^{\overline{n}} = 0$ for $n \geq 1$, such a term should better not appear in the denominator.

For $\mathbb{K} = \mathbb{C}$ we can borrow another extension of the factorial from the theory of special functions: the gamma function. Following Euler, we first define

$$L(z) := \lim_{k \to \infty} \binom{k+z-1}{k} k^{1-z} \qquad (z \in \mathbb{C}).$$

This limit exists for all $z \in \mathbb{C}$ and we have $L(z) = 0$ if and only if $z \in \{0, -1, -2, \dots\}$. For $z \in \mathbb{C} \setminus \{0, -1, -2, \dots\}$ we can therefore define $\Gamma(z) := 1/L(z)$. It can be shown that Γ is analytic in its domain of definition, and that it has simple poles at $z = 0, -1, -2, \dots$. Furthermore, we have functional equations such as

$$\Gamma(z+1) = z\Gamma(z) \qquad (z \in \mathbb{C} \setminus \{0, -1, -2, \dots\}),$$

$$\Gamma(1-z)\Gamma(z) = \frac{\pi}{\sin(\pi z)} \qquad (z \in \mathbb{C} \setminus \mathbb{Z}),$$

and some special values such as $\Gamma(1) = 1$ or $\Gamma(\frac{1}{2}) = \sqrt{\pi}$ or $\Gamma'(1) = -\gamma$. For proofs and additional properties we refer to textbooks on special functions like [7]. What matters for us is mostly that we can use the gamma function to express hypergeometric sequences in closed form. If $r(x) \in \mathbb{C}(x)$ is any rational function, say

$$r(x) = c \frac{(x+z_1)(x+z_2)\cdots(x+z_p)}{(x+w_1)(x+w_2)\cdots(x+w_q)}$$

for some $c \in \mathbb{C}$ and $z_1, \dots, z_p, w_1, \dots, w_q \in \mathbb{C}$, then for

$$a_n := bc^n \frac{\Gamma(n+z_1)\Gamma(n+z_2)\cdots\Gamma(n+z_p)}{\Gamma(n+w_1)\Gamma(n+w_2)\cdots\Gamma(n+w_q)}$$

with $b \in \mathbb{C}$ fixed we will have $a_{n+1} = r(n)a_n$ for all $n \in \mathbb{N}$ where a_n, a_{n+1} and $r(n)$ are defined.

This way of expressing a hypergeometric sequence is nice and natural as long as we do not run into one of the singularities of the gamma function, viz. as long as $z_i, w_j \notin \{0, -1, -2, \dots\}$ for all z_i, w_j. But also when there are singularities, we might not be lost completely. Since Γ is a continuous function, singularities may be removable by way of continuous extension, just like the function $z \mapsto \frac{z-1}{z^2-1}$ admits a continuous extension to the point $z = 1$ where it is a priori undefined. For example, the function

$$z \mapsto \frac{1}{\Gamma(z-10)},$$

which is undefined for $z \in \{\ldots, 8, 9, 10\}$, can be extended continuously to all complex numbers by setting

$$f: \mathbb{C} \to \mathbb{C}, \quad f(z) := \begin{cases} 0 & \text{if } z \in \{\ldots, 8, 9, 10\} \\ 1/\Gamma(z - 10) & \text{otherwise} \end{cases}.$$

We then have $f(z+1) = \frac{1}{z-10} f(z)$ for all $z \in \mathbb{C}$, and in particular $(f(n))_{n=0}^{\infty}$ is a sequence in which all terms are defined. In contrast, the function $z \mapsto \Gamma(z - 10)$, which is also undefined for $z \in \{\ldots, 8, 9, 10\}$, cannot be extended continuously to a function $f: \mathbb{C} \to \mathbb{C}$ satisfying $f(z+1) = (z - 10)f(z)$ everywhere. Its singularities are poles. In order to construct a function with shift quotient $z - 10$, we can take the function

$$z \mapsto \frac{\cos(\pi z)}{\Gamma(11 - z)},$$

which is a priori undefined for $z \in \{\ldots, 11, 12, 13\}$, and extend it continuously to

$$f: \mathbb{C} \to \mathbb{C}, \quad f(z) := \begin{cases} 0 & \text{if } z \in \{11, 12, 13, \ldots\} \\ \cos(\pi z)/\Gamma(11 - z) & \text{otherwise} \end{cases}.$$

We then have $f(z+1) = (z - 10)f(z)$ for all $z \in \mathbb{C}$, and in particular $(f(n))_{n=0}^{\infty}$ is a sequence in which all terms are defined. For an arbitrary rational function $r(x) \in \mathbb{C}(x)$ we will always find a complex function f which is continuously extensible to all points in $\mathbb{N}$ (at least) and satisfies $f(n+1) = r(n)f(n)$ for all $n \in \mathbb{N}$ where $r(n)$ is defined. From now on, we adopt the convention that we will remove all removable singularities without notice, so that calculations (and in particular: cancellations) with expressions involving the gamma function can be performed without being too concerned about singular points. Only particularly nasty situations might require us to explicitly calculate a limit.

Let us now turn to closure properties for hypergeometric sequences. It follows directly from the definition that if $(a_n)_{n=0}^{\infty}$ and $(b_n)_{n=0}^{\infty}$ are hypergeometric, then so are their Hadamard product $(a_n b_n)_{n=0}^{\infty}$ and the reciprocal $(1/a_n)_{n=0}^{\infty}$ (if $a_n \neq 0$ for all n). Also the dilation $(a_{un+v})_{n=0}^{\infty}$ ($u, v \in \mathbb{N}$ fixed) of a hypergeometric sequence is hypergeometric. But the sum $(a_n + b_n)_{n=0}^{\infty}$ is in general not hypergeometric. For, suppose that $(a_n)_{n=0}^{\infty}$ and $(b_n)_{n=0}^{\infty}$ are such that

$$a_{n+1} = r(n)a_n \quad \text{and} \quad b_{n+1} = s(n)b_n$$

for some rational functions $r(x), s(x) \in \mathbb{K}(x)$. If there is another rational function $t(x) \in \mathbb{K}(x)$ with

$$(a_{n+1} + b_{n+1}) = t(n)(a_n + b_n)$$

then we will have

$$r(n)a_n + s(n)b_n = t(n)(a_n + b_n).$$

This forces the relation

$$a_n = \frac{t(n) - s(n)}{r(n) - t(n)} b_n,$$

i.e., we can get from one sequence to the other by multiplying with a certain rational function. In general, this is not going to happen (think of $a_n = n!$ and $b_n = 1$), but it may happen (think of $a_n = n!$ and $b_n = (n+1)!$). If two hypergeometric sequences $(a_n)_{n=0}^{\infty}$ and $(b_n)_{n=0}^{\infty}$ in $\mathbb{K}$ are such that $p(n)a_n = q(n)b_n$ for some polynomials $p(x), q(x) \in \mathbb{K}[x]$ and all $n \in \mathbb{N}$, then $(a_n)_{n=0}^{\infty}$ and $(b_n)_{n=0}^{\infty}$ are called *similar*. According to the calculation above, two hypergeometric sequences are similar if and only if their sum is hypergeometric. (Strictly speaking, this equivalence only holds if we also allow rational functions whose numerator and denominator have factors in common.) As a particular special case, it is clear that a hypergeometric sequence $(a_n)_{n=0}^{\infty}$ is similar to all of its shifted versions $(a_{n+i})_{n=0}^{\infty}$ ($i \in \mathbb{N}$ fixed).

Sequences in several variables are called hypergeometric if they are hypergeometric with respect to each of their arguments. In particular, a bivariate sequence $(a_{n,k})_{n,k=0}^{\infty}$ is called hypergeometric if there exist rational functions $u(x,y)$ and $v(x,y) \in \mathbb{K}(x,y)$ such that

$$a_{n+1,k} = u(n,k)a_{n,k} \quad \text{and} \quad a_{n,k+1} = v(n,k)a_{n,k}$$

for all $n, k \in \mathbb{N}$ where $u(n,k)$ and $v(n,k)$ are defined. For example, by

$$\binom{n+1}{k} = \frac{n+1}{n-k+1}\binom{n}{k} \quad \text{and} \quad \binom{n}{k+1} = \frac{n-k}{k+1}\binom{n}{k},$$

the binomial coefficients are hypergeometric.

In contrast to the univariate case, not for every choice of rational functions $u(x,y)$ and $v(x,y) \in \mathbb{K}(x,y)$ there exists a (nontrivial) bivariate sequence $(a_{n,k})_{n,k=0}^{\infty}$ which has these rational functions as its shift quotients. The rational functions must satisfy a certain compatibility condition which originates from the need that a shift in n followed by a shift in k should give the same as a shift in k followed by a shift in n: In order to have both

$$a_{n+1,k+1} = u(n,k+1)a_{n,k+1} = u(n,k+1)v(n,k)a_{n,k}$$
$$\text{and} \quad a_{n+1,k+1} = v(n+1,k)a_{n+1,k} = u(n,k)v(n+1,k)a_{n,k},$$

the rational functions $u(x,y)$ and $v(x,y)$ must be such that

$$\frac{u(x,y+1)}{u(x,y)} = \frac{v(x+1,y)}{v(x,y)}.$$

If this condition is satisfied, then there exists a bivariate hypergeometric sequence which has $u(x,y)$ and $v(x,y)$ as shift quotients.

However, in the case $\mathbb{K} = \mathbb{C}$, this does not imply that a bivariate hypergeometric sequence can be expressed in terms of the gamma function. For example, no such representation is available for the sequence $(1/(n^2+k^2+1))_{n,k=0}^{\infty}$, although it is surely hypergeometric. Those bivariate hypergeometric sequences which can be expressed in terms of the gamma function are called proper hypergeometric. Precisely, we say that a sequence $(a_{n,k})_{n,k=0}^{\infty}$ in $\mathbb{C}$ is called *proper hypergeometric*, if there exists a polynomial $b(x,y) \in \mathbb{C}[x,y]$ and constants $c,d \in \mathbb{C}, r_i,s_i,e_i \in \mathbb{Z}, t_i \in \mathbb{C}$ $(i=1,\ldots,m)$ such that

$$a_{n,k} = c^k d^n \lim_{\varepsilon \to 0} \left(\lim_{\delta \to 0} f(n+\varepsilon,k+\delta) \right) \qquad (n,k \ge 0)$$

where $f(z,w) = b(z,w)\Gamma(r_1z+s_1w+t_1)^{e_1} \cdots \Gamma(r_mz+s_mw+t_m)^{e_m}$.

The limits indicate that we again remove singularities by continuous extensions where this is possible. For example, the bivariate sequence of binomial coefficients is proper hypergeometric because we have

$$\binom{n}{k} = \lim_{\varepsilon \to 0} \lim_{\delta \to 0} \frac{\Gamma((n+\varepsilon)+1)}{\Gamma((k+\delta)+1)\Gamma((n+\varepsilon)-(k+\delta)+1)} \qquad (n \in \mathbb{C}, k \in \mathbb{Z}).$$

We can now also give a reasonable definition of $\binom{n}{k}$ when $k \notin \mathbb{Z}$ by defining $\binom{n}{k}$ as the value of this limit also for these k.

5.3 The Tetrahedron for Hypergeometric Sequences

By rephrasing the definition, a sequence $(a_n)_{n=0}^{\infty}$ in $\mathbb{K}$ is hypergeometric if and only if it satisfies a first order linear recurrence equation with polynomial coefficients,

$$p(n)a_{n+1} - q(n)a_n = 0$$

for some $p(x),q(x) \in \mathbb{K}[x]$ and all $n \in \mathbb{N}$ with $p(n) \neq 0$. If $p(x)$ has no roots in $\mathbb{N}$, then this recurrence together with a single initial value $a_0 \in \mathbb{K}$ uniquely determines all the sequence terms a_n $(n \ge 0)$. In general, if $p(n) = 0$ for some $n \in \mathbb{N}$, then the recurrence does not tell us anything about the value a_{n+1}. Since $p(x)$ is a polynomial, there can be only finitely many such points $n \in \mathbb{N}$. They are called the *singularities* of the recurrence. If $n_1,\ldots,n_s \in \mathbb{N}$ are the singularities of the recurrence, then the recurrence determines all the terms of a solution $(a_n)_{n=0}^{\infty}$ except for the initial term a_0 and the terms $a_{n_1+1}, a_{n_2+1}, \ldots, a_{n_s+1}$. Consequently, the solutions of a first order linear recurrence with polynomial coefficients form a $\mathbb{K}$-vector space of dimension $s+1$, where s is the number of singularities.

We now turn from the recurrence vertex of the Tetrahedron to the asymptotics vertex. For hypergeometric sequences in $\mathbb{C}$ which admit a representation in terms of the gamma function, the asymptotic behavior is immediate from the asymptotic behavior of the gamma function, which is given by the formula

$$\Gamma(n+z) \sim \sqrt{2\pi}\, n^{z+n-\frac{1}{2}} e^{-n} \qquad (n \to \infty, n \in \mathbb{N}, z \in \mathbb{C}).$$

Two special cases of this formula are of particular interest. The first is the case $z = 1$ which gives Stirling's celebrated formula on the asymptotics of $(n!)_{n=0}^{\infty}$. The second is the formula

$$\frac{a^{\overline{n}}}{b^{\overline{n}}} \sim \frac{\Gamma(b)}{\Gamma(a)} n^{a-b} \qquad (n \to \infty)$$

which is valid for all $a, b \in \mathbb{C} \setminus \{0, -1, -2, \dots\}$ and which follows directly from the general formula by $z^{\overline{n}} = \Gamma(n+z)/\Gamma(z)$. This formula for the quotient of two rising factorials is often useful in practice because rising factorials have a tendency to occur in fractions with balanced numerators and denominators, like in the example sequence $(a_n)_{n=0}^{\infty}$ defined via

$$2(n+2)(n+1)a_{n+1} - (2n+1)^2 a_n = 0 \quad (n \geq 0), \qquad a_0 = 1.$$

For this sequence we have

$$a_n = 2^n \frac{(1/2)^{\overline{n}}(1/2)^{\overline{n}}}{1^{\overline{n}} 2^{\overline{n}}} \sim 2^n \frac{\Gamma(1)\Gamma(2)}{\Gamma(1/2)\Gamma(1/2)} n^{\frac{1}{2}+\frac{1}{2}-1-2} = \frac{2^n}{\pi n^2} \qquad (n \to \infty).$$

Just like in this example, determining the asymptotic behavior of a hypergeometric sequence reduces to simple rewriting if the asymptotic formula for $\Gamma(n+z)$ is taken for granted. A formal derivation of that formula, however, requires some work and we will not reproduce it here because this would lead us too far astray. Instead, let us only give a somewhat informal argument why the asymptotic estimate is plausible in the case $z = 1$, Stirling's formula. The task is then to find a sequence $(a_n)_{n=0}^{\infty}$ such that $\lim_{n \to \infty} n!/a_n = 1$. Because of

$$\frac{(n+1)!}{n!} = n+1 \qquad (n \geq 0)$$

we better take a sequence $(a_n)_{n=0}^{\infty}$ with $a_{n+1}/a_n \sim n+1$ $(n \to \infty)$ for otherwise the quotient $n!/a_n$ has no chance to converge to 1. For the choice $a_n = n^n$ we have

$$\frac{a_{n+1}}{a_n} = (n+1)\left(1+\frac{1}{n}\right)^n \sim (n+1)e \qquad (n \to \infty)$$

which is not exactly what we want, but pretty close. We need a correction term for getting rid of the unwanted constant factor e. If we set $a_n = e^{-n} n^n$, then

$$\frac{a_{n+1}}{a_n} = e^{-1}(n+1)\left(1+\frac{1}{n}\right)^n \sim n+1 \qquad (n \to \infty),$$

as needed. At this point we can record that $n!/a_n = O(1)$ $(n \to \infty)$. We are almost there. Assume next that we have $n!/a_n \sim c n^{\alpha}$ $(n \to \infty)$ for some $\alpha \leq 0$ and $c \in \mathbb{R} \setminus \{0\}$. Comparing again the asymptotics of the shift quotients, we have by the binomial theorem

$$\frac{c(n+1)^{\alpha}}{c n^{\alpha}} = \left(1+\frac{1}{n}\right)^{\alpha} = 1 + \alpha n^{-1} + \binom{\alpha}{2} n^{-2} + \cdots \qquad (n \to \infty)$$

on one hand, and on the other hand

$$
\frac{(n+1)!/a_{n+1}}{n!/a_n} \sim e^{-1}\left(1+\frac{1}{n}\right)^n = e^{-1}\exp\left(n\log\left(1+\frac{1}{n}\right)\right)
$$

$$
= e^{-1}\exp\left(n\left(n^{-1}-\tfrac{1}{2}n^{-2}+\tfrac{1}{3}n^{-3}-\tfrac{1}{4}n^{-4}+\cdots\right)\right)
$$

$$
= \exp\left(-\tfrac{1}{2}n^{-1}+\tfrac{1}{3}n^{-2}+\cdots\right)
$$

$$
= 1+\left(-\tfrac{1}{2}n^{-1}+\cdots\right)+\tfrac{1}{2}\left(-\tfrac{1}{2}n^{-1}+\cdots\right)^2+\cdots \quad (n\to\infty).
$$

Comparing the coefficients of n^{-1} in both expansions yields $\alpha = -\tfrac{1}{2}$. Indeed, with the once more refined definition $a_n := n^{-1/2}e^{-n}n^n$ we have $\lim_{n\to\infty} n!/a_n = c$ for some nonzero constant $c \in \mathbb{R}$. What is finally this constant? This is a notoriously hard question in asymptotics, and it is often not possible to give a satisfactory answer. But using convergence acceleration techniques such as the Richardson method from Problem 2.19, we can at least find good approximations to the constant, for example,

$$
\lim_{n\to\infty} \frac{n!}{n^{n-1/2}e^{-n}} \approx 2.506628274631087458\ldots
$$

and this matches the expected $\sqrt{2\pi} = 2.50662827463100005024\ldots$ fairly accurately. Of course, this does not prove anything, but it gives striking empirical evidence in support of the correctness of Stirling's formula. Readers interested in rigorous arguments are referred to [24, 7, 21] where several independent proofs can be found.

Let us now turn to generating functions. What are the generating functions of hypergeometric sequences? Well, there is no better way of describing them than saying that they are the formal power series whose coefficient sequences are hypergeometric sequences. We call such power series *hypergeometric series* and introduce the somewhat intimidating but widely accepted notation

$$
{}_pF_q\left(\begin{matrix} a_1, a_2, \ldots, a_p \\ b_1, \ldots, b_q \end{matrix}\,\middle|\, x\right) := \sum_{k=0}^{\infty} \frac{a_1^{\overline{k}} a_2^{\overline{k}} \cdots a_p^{\overline{k}}}{b_1^{\overline{k}} \cdots b_q^{\overline{k}}} \frac{x^k}{k!} \in \mathbb{K}[[x]]
$$

for $a_1, \ldots, a_p \in \mathbb{K}$ and $b_1, \ldots, b_q \in \mathbb{K} \setminus \{0, -1, -2, \ldots\}$. We do not need to take into account possible exponential factors c^k, because they are covered by the substitution $x \mapsto cx$. Also the requirement that there be a factor $k!$ in the denominator does not restrict the generality, because if we want to get rid of this factor, we can just introduce an additional parameter $a_{p+1} = 1$ and the new factor $1^{\overline{k}} = k!$ in the numerator will cancel the $k!$ in the denominator. For $\mathbb{K} = \mathbb{C}$, we can use $a^{\overline{k}} = \Gamma(k+a)/\Gamma(a)$ to rewrite the expression for the summand into the form given earlier for hypergeometric sequences in $\mathbb{C}$.

Many elementary series can be expressed in terms of hypergeometric series. For example,

$$_0F_0\left(\begin{matrix}-\\-\end{matrix}\middle|x\right)=\exp(x),\qquad\qquad _1F_0\left(\begin{matrix}a\\-\end{matrix}\middle|x\right)=\frac{1}{(1-x)^a},$$

$$_2F_1\left(\begin{matrix}1,1\\2\end{matrix}\middle|-x\right)=\frac{1}{x}\log(1+x),\qquad _2F_1\left(\begin{matrix}-n,n\\\frac{1}{2}\end{matrix}\middle|\frac{1-x}{2}\right)=T_n(x)\ (n\ge 0),$$

$$_0F_1\left(\begin{matrix}-\\\frac{3}{2}\end{matrix}\middle|-\frac{x^2}{4}\right)=\frac{1}{x}\sin(x),\qquad _2F_1\left(\begin{matrix}\frac{1}{2},1\\\frac{3}{2}\end{matrix}\middle|-x^2\right)=\frac{1}{x}\arctan(x).$$

The series for the Chebyshev polynomial $T_n(x)$ is noteworthy: the parameter $a_1 = -n$ corresponds to a factor $(-n)^{\overline{k}}$ in the coefficient sequence and this factor (and hence the entire coefficient) is zero if $k > n$. This truncates the infinite series to a polynomial, and this is why the substitution $x \mapsto \frac{1-x}{2}$ is meaningful in this case.

Identities about hypergeometric sequences correspond to identities about hypergeometric series, and working on one of these two equivalent levels may at times be easier than working on the other. So it is worthwhile to build up a knowledge base with relations about the hypergeometric series. The systematic search for general identities connecting hypergeometric series to each other is going on for now more than two hundred years, and although a long list of such relations is available, it still happens that new ones are discovered. Among the most basic (and the most important) relations, there are

$$_2F_1\left(\begin{matrix}-n,a\\c\end{matrix}\middle|1\right)=\frac{(c-a)^{\overline{n}}}{c^{\overline{n}}}\qquad\qquad (n\ge 0)\quad\text{(Vandermonde)},$$

$$_3F_2\left(\begin{matrix}-n,a,b\\c,1+a+b-c-n\end{matrix}\middle|1\right)=\frac{(c-a)^{\overline{n}}(c-b)^{\overline{n}}}{c^{\overline{n}}(c-a-b)^{\overline{n}}}\qquad (n\ge 0)\quad\text{(Pfaff-Saalschütz)},$$

$$_2F_1\left(\begin{matrix}a,b\\c\end{matrix}\middle|x\right)=(1-x)^{c-a-b}\,_2F_1\left(\begin{matrix}c-a,c-b\\c\end{matrix}\middle|x\right)\qquad\text{(Euler; Ex. 5.6)},$$

$$_2F_1\left(\begin{matrix}a,b\\c\end{matrix}\middle|x\right)=(1-x)^{-a}\,_2F_1\left(\begin{matrix}a,c-b\\c\end{matrix}\middle|\frac{x}{x-1}\right)\qquad\text{(Pfaff; Ex. 5.6)},$$

$$_2F_1\left(\begin{matrix}a,b\\2b\end{matrix}\middle|\frac{4x}{(1+x)^2}\right)=(1+x)^{2a}\,_2F_1\left(\begin{matrix}a,a-b+\frac{1}{2}\\b+\frac{1}{2}\end{matrix}\middle|x^2\right)\qquad\text{(Landen; Ex. 7.11)},$$

$$_2F_1\left(\begin{matrix}a,b\\c\end{matrix}\middle|x\right)=\,_2F_1\left(\begin{matrix}a+1,b\\c\end{matrix}\middle|x\right)-\frac{bx}{c}\,_2F_1\left(\begin{matrix}a+1,b+1\\c+1\end{matrix}\middle|x\right)\qquad\text{(Gauss; Ex. 5.7)}.$$

Lots of further items can be found in the literature [7]. Note that the substitution $x \mapsto 1$ in the first two identities is legitimate because the integer parameter $-n$ ensures that these series are just polynomials. All these identities are true in the sense of formal power series. There are also identities which are only meaningful analytically, for example Gauss' summation formula

$$_2F_1\left(\begin{matrix} a,b \\ c \end{matrix}\middle| 1\right) = \frac{\Gamma(c)\Gamma(c-a-b)}{\Gamma(c-a)\Gamma(c-b)}$$

which holds for all $a,b,c \in \mathbb{C}$ where both sides are defined and $\mathrm{Re}(c-a-b) > 0$. But we will not consider such identities.

As an example application of hypergeometric transformations, consider the sum

$$s_n := \sum_{k=0}^{n} (-1)^k \binom{m}{k} \qquad (n \geq 0)$$

where m is a formal parameter. Let $a_k := (-1)^k \binom{m}{k} = \frac{(-m)^{\overline{k}}}{k!}$ be the summand sequence. In order to rephrase the sum s_n as a hypergeometric series we need to introduce a factor which truncates the sum after the n-th term. It would be useful to have a factor $(-n)^{\overline{k}}$ in the summand which would turn it to 0 for $k \geq 0$, but there is none. If we artificially introduce one, then we must compensate for it by dividing it out again. But, once again, $(-n)^{\overline{k}}$ is zero for $k \geq 0$, so it might not be a good idea to have such a term in a denominator. To avoid a division by zero, we disturb the denominator slightly. The reformulation

$$a_k = \lim_{\varepsilon \to 0} \frac{(-n)^{\overline{k}}(-m)^{\overline{k}}}{(-n+\varepsilon)^{\overline{k}}k!} \qquad (k \geq 0)$$

then implies the representation

$$s_n = \lim_{\varepsilon \to 0} {}_2F_1\left(\begin{matrix} -n,-m \\ -n+\varepsilon \end{matrix}\middle| 1\right) \qquad (n \geq 0).$$

Now we can invoke the $_2F_1$-version of Vandermonde's summation formula to obtain

$$s_n = \lim_{\varepsilon \to 0} \frac{(m-n+\varepsilon)^{\overline{n}}}{(-n+\varepsilon)^{\overline{n}}} = \frac{(m-n)(m-n+1)\cdots(m-1)}{(-n)(-n+1)\cdots(-1)} = (-1)^n \binom{m-1}{n}$$

for $n \geq 0$. This example brings us to the last remaining corner of the Tetrahedron: summation. We consider this algorithmically most exciting corner separately in the following two sections.

5.4 Indefinite Summation

If $(a_n)_{n=0}^{\infty}$ is a hypergeometric sequence, the sequence $(\sum_{k=0}^{n} a_k)_{n=0}^{\infty}$ need not be hypergeometric again. For example, while $(1/n)_{n=1}^{\infty}$ is clearly hypergeometric, the sequence $(H_n)_{n=0}^{\infty}$ with $H_n = \sum_{k=1}^{n} \frac{1}{k}$ is clearly not, because we have seen that $H_n \sim$

$\log(n)$ $(n \to \infty)$, and this is not compatible with what we have found before for the possible asymptotics of hypergeometric sequences. On the other hand, the identity

$$\sum_{k=0}^{n} kk! = (n+1)! - 1 \qquad (n \geq 0)$$

is an example for a hypergeometric sum that admits a hypergeometric closed form.

Our goal is to understand under which circumstances a hypergeometric closed form exists. To this end, let $(a_n)_{n=0}^{\infty}$ be a hypergeometric sequence in $\mathbb{K}$ with

$$a_{n+1} = u(n)a_n$$

for some rational function $u(x) \in \mathbb{K}(x)$ and all $n \in \mathbb{N}$ where $u(n)$ is defined. Let us assume that $a_n \neq 0$ for infinitely many $n \in \mathbb{N}$. (Otherwise the summation problem is trivial anyway.) We say that a hypergeometric closed form for $\sum_{k=0}^{n} a_k$ exists if there is a hypergeometric solution $(s_n)_{n=0}^{\infty}$ of the telescoping equation

$$s_{n+1} - s_n = a_n \qquad (n \geq 0),$$

because then we have $\sum_{k=0}^{n} a_k = s_{n+1} - s_0$ $(n \geq 0)$. Suppose such a hypergeometric solution $(s_n)_{n=0}^{\infty}$ exists, say $s_{n+1} = v(n)s_n$ for some rational function $v(x) \in \mathbb{K}(x)$ and all $n \in \mathbb{N}$ where $v(n)$ is defined. Then we have

$$(v(n) - 1)s_n = a_n$$

for all these n. This means that $(s_n)_{n=0}^{\infty}$ must be similar to $(a_n)_{n=0}^{\infty}$, i.e., we have $s_n = w(n)a_n$ for some rational function $w(x) \in \mathbb{K}(x)$ and all $n \in \mathbb{N}$ where $w(n)$ is defined and nonzero. The telescoping equation for $(s_n)_{n=0}^{\infty}$ implies that $w(x)$ must satisfy

$$u(n)w(n+1) - w(n) = 1$$

for all $n \in \mathbb{N}$ where $a_n \neq 0$ and $u(n)$ is defined and both $w(n)$ and $w(n+1)$ are defined and nonzero.

Keeping track of all those conditions on exceptional points is getting clumsy and starts to blur the overall picture. To improve readability, let us introduce the set $\mathbf{N} \subseteq \mathbb{N}$ of all $n \in \mathbb{N}$ where $u(n)$ is defined and both $w(n)$ and $w(n+1)$ are defined and nonzero. Observe that $\mathbb{N} \setminus \mathbf{N}$ is always finite. Also observe that $\mathbf{N}$ depends on the input $u(x)$ as well as on the output $w(x)$. Readers not interested in discussing exceptional points may read "$n \in \mathbf{N}$" simply as "all but finitely many $n \in \mathbb{N}$".

We have seen that $u(n)w(n+1) - w(n) = 1$ ($n \in \mathbf{N}$ with $a_n \neq 0$) is a necessary condition on the rational function $w(x)$. It is also a sufficient condition in the sense that if $w(x)$ is such that $u(n)w(n+1) - w(n) = 1$ ($n \in \mathbf{N}$ with $a_n \neq 0$) then for any hypergeometric sequence $(s_n)_{n=0}^{\infty}$ with $s_n = w(n)a_n$ ($n \in \mathbf{N}$ with $w(n)$ defined and nonzero) we have $s_{n+1} - s_n = a_n$ ($n \in \mathbf{N}$). This is the best we can get; the finitely many points $n \in \mathbb{N} \setminus \mathbf{N}$ must be inspected separately.

In order to solve the summation problem, it is therefore necessary and sufficient to determine rational functions $w(x) \in \mathbb{K}(x)$ with

$$u(n)w(n+1) - w(n) = 1 \qquad (n \in \mathbf{N} \text{ with } a_n \neq 0).$$

As this equation is supposed to hold for infinitely many points n, it holds if and only if the equation

$$u(x)w(x+1) - w(x) = 1$$

holds in $\mathbb{K}(x)$. For the two sides of that equation to be equal, the numerators and denominators on both sides must be equal. Let us look at the denominator first. If we write $u(x) = u_1(x)/u_2(x)$ and $w(x) = w_1(x)/w_2(x)$ for relatively prime polynomials $u_1(x), u_2(x)$ and $w_1(x), w_2(x) \in \mathbb{K}[x]$ then the equation becomes

$$\frac{u_1(x)w_1(x+1)w_2(x) - u_2(x)w_2(x+1)w_1(x)}{u_2(x)w_2(x+1)w_2(x)} = 1.$$

This can only become true if appropriate cancellation happens on the left hand side. In particular, we must have

$$w_2(x) \mid u_1(x)w_1(x+1)w_2(x) - u_2(x)w_2(x+1)w_1(x)$$
$$\implies w_2(x) \mid u_2(x)w_2(x+1)w_1(x)$$
$$\implies w_2(x) \mid u_2(x)w_2(x+1)$$

and similarly

$$w_2(x+1) \mid u_1(x)w_1(x+1)w_2(x) - u_2(x)w_2(x+1)w_1(x)$$
$$\implies w_2(x+1) \mid u_1(x)w_1(x+1)w_2(x)$$
$$\implies w_2(x+1) \mid u_1(x)w_2(x)$$
$$\implies w_2(x) \mid u_1(x-1)w_2(x-1).$$

But this does not leave too much freedom for $w_2(x)$ after all. If $p(x) \in \mathbb{K}[x]$ is an irreducible factor of $w_2(x)$, then we must have either $p(x) \mid u_1(x-1)$ or $p(x) \mid w_2(x-1)$. In the latter case, we have $p(x+1) \mid w_2(x)$, which, by the same argument, requires either $p(x+1) \mid u_1(x-1)$ or $p(x+1) \mid w_2(x-1)$. The argument cannot repeat indefinitely because $w_2(x)$ is a polynomial of finite degree which cannot have infinitely many irreducible factors $p(x), p(x+1), p(x+2)$, etc., so for some $i \in \mathbf{N}$ we must eventually have $p(x+i) \mid u_1(x-1)$. We can play the same game with the constraint $w_2(x) \mid u_2(x)w_2(x+1)$ and find that for some $j \in \mathbf{N}$ we must have $p(x-j) \mid u_2(x)$.

Putting both observations together, we find that a solution $w(x)$ can only have a non-trivial denominator if there is a common factor between $u_1(x)$ and $u_2(x+i)$ for some suitable $i \in \mathbf{N} \setminus \{0\}$. If we could make a substitution to our equation that will assure that $u_1(x)$ and $u_2(x+i)$ have no common factors for any $i \in \mathbf{N}$, then we could conclude that $w_2(x) = 1$ and we would not need to worry any further about the denominator. This is the point where Gosper enters the stage. Following his advice, we construct polynomials $p(x), q(x), r(x) \in \mathbb{K}[x]$ such that

$$u(x) = \frac{p(x+1)}{p(x)} \frac{q(x)}{r(x+1)} \quad \text{and} \quad \gcd(q(x), r(x+i)) = 1 \quad (i \in \mathbb{N} \setminus \{0\}).$$

Such a representation is called a *Gosper form* of $u(x)$. We will see in Problem 5.8 that every rational function $u(x)$ can be brought to this form. Assume for the moment that we have found $p(x), q(x), r(x) \in \mathbb{K}[x]$ with the desired property. Plugging this representation into our equation

$$u(x)w(x+1) - w(x) = 1$$

and multiplying by $p(x)$ produces

$$\frac{q(x)}{r(x+1)} p(x+1)w(x+1) - p(x)w(x) = p(x).$$

Now substitute $\bar{w}(x) = p(x)w(x)$ to obtain the new equation

$$\frac{q(x)}{r(x+1)} \bar{w}(x+1) - \bar{w}(x) = p(x).$$

We seek a rational function solution $\bar{w}(x)$ in $\mathbb{K}(x)$ to this equation. Such a solution, however, cannot have a nontrivial denominator, because we have seen before (with $u_1(x)$ in place of $q(x)$ and $u_2(x)$ in place of $r(x+1)$) that any nontrivial factor of the denominator must be a common factor between $q(x)$ and some shift $r(x+i)$, which would be in conflict with the condition in the Gosper form. It is therefore enough to search for polynomial solutions $\bar{w}(x)$ of the new equation. Furthermore, because also the denominator $r(x+1)$ must cancel out, we must have $r(x+1) \mid \bar{w}(x+1)$ for any solution $\bar{w}(x)$, i.e., $r(x) \mid \bar{w}(x)$ and we can do another substitution $\bar{w}(x) = r(x)y(x)$ to finally obtain

$$q(x)y(x+1) - r(x)y(x) = p(x).$$

This is called the *Gosper equation*. Since we know $p(x), q(x), r(x) \in \mathbb{K}[x]$ explicitly, we can find all its polynomial solutions $y(x)$ with the algorithm described in Sect. 3.4.

By finally setting $w(x) = \frac{r(x)}{p(x)}y(x) \in \mathbb{K}(x)$, we find that any polynomial solution $y(x) \in \mathbb{K}[x]$ of the Gosper equation gives rise to a telescoping relation

$$s_{n+1} - s_n = a_n \quad (n \in \mathbb{N})$$

for $s_n := w(n)a_n$ ($n \in \mathbb{N}$ where $w(n)$ is defined and nonzero). Conversely, we have also shown that any hypergeometric solution $(s_n)_{n=0}^{\infty}$ of the telescoping equation implies the existence of a polynomial solution $y(x) \in \mathbb{K}[x]$ of the Gosper equation. In other words, the telescoping equation has a hypergeometric solution if and only if the Gosper equation has a polynomial solution.

Putting things together, we obtain *Gosper's algorithm* for solving the telescoping equation: It takes as input a rational function $u(x) \in \mathbb{K}(x)$ with $a_{n+1} = u(n)a_n$ for all $n \in \mathbb{N}$ where $u(n)$ is defined. As output it returns either rational functions $w(x) \in \mathbb{K}(x)$ such that for every hypergeometric sequence $(s_n)_{n=0}^{\infty}$ with $s_n = w(n)a_n$ ($n \in \mathbb{N}$ where $w(n)$ is defined and nonzero) we have $s_{n+1} - s_n = a_n$ ($n \in \mathbb{N}$), or it returns the message "no hypergeometric solution" if no hypergeometric solution $(s_n)_{n=0}^{\infty}$ exists. The algorithm consists of the following steps:

1. Find $p(x), q(x), r(x) \in \mathbb{K}[x]$ such that

$$u(x) = \frac{p(x+1)}{p(x)} \frac{q(x)}{r(x+1)} \quad \text{and} \quad \gcd(q(x), r(x+i)) = 1 \quad (i \in \mathbb{N} \setminus \{0\})$$

   (using for example the algorithm from Problem 5.8).

2. Solve the equation

$$p(x) = q(x)y(x+1) - r(x)y(x)$$

   for polynomials $y(x) \in \mathbb{K}[x]$ (using for example the algorithm from Sect. 3.4).

3. If no polynomial solution exists, then return "no hypergeometric solution" and stop. Otherwise, return $w(x) = \frac{r(x)}{p(x)} y(x)$.

The output description can be simplified if $u(n)$ is defined for all $n \in \mathbb{N}$. There are no exceptional points in this case: if the algorithm returns a rational function $w(x) \in \mathbb{K}(x)$ then $w(n)$ will be defined for all $n \in \mathbb{N}$ and a hypergeometric solution $(s_n)_{n=0}^{\infty}$ is given by $s_n := w(n)a_n$ ($n \in \mathbb{N}$). In particular, we then have

$$\sum_{k=0}^{n} a_k = s_{n+1} - s_0 \quad (n \geq 0).$$

Now it is time for an example. Consider again the sum

$$\sum_{k=0}^{n} (-1)^k \binom{m}{k} \quad (n \geq 0)$$

where m is a formal parameter. For $a_k = (-1)^k \binom{m}{k}$ we have

$$a_{k+1} = \frac{k-m}{k+1} a_k \quad (k \geq 0),$$

so $u(x) = \frac{x-m}{x+1}$ in this case. Note that $a_k \neq 0$ for all $k \in \mathbb{N}$ and $u(k)$ is defined for all $k \in \mathbb{N}$. We can see by inspection that no positive integer shift of the denominator of $u(x)$ will ever match the numerator, so we can safely choose $p(x) = 1$, $q(x) = x - m$ and $r(x) = x$. For the Gosper equation

$$1 = (x-m)y(x+1) - xy(x)$$

we find the solution $y(x) = -\frac{1}{m} \in \mathbb{Q}(m)[x]$. Hence $s_n = -\frac{n}{m}(-1)^n \binom{m}{n}$ solves the telescoping equation $s_{n+1} - s_n = a_n$ for all $n \in \mathbb{N}$. As a consequence, we find the closed form evaluation

$$\sum_{k=0}^{n}(-1)^k\binom{m}{k} = s_{n+1} - s_0 = \frac{n+1}{m}(-1)^n\binom{m}{n+1} - 0 = (-1)^n\binom{m-1}{n} \quad (n \geq 0).$$

5.5 Definite Summation

If we apply Gosper's algorithm to the sum

$$\sum_{k=0}^{n}\binom{m}{k} \quad (n \geq 0),$$

where m is a formal parameter, then it will tell us that this sum does not have a hypergeometric closed form. While this is a plausible answer as long as n and m are unrelated, we know that in the special case $n = m$ the sum exceptionally simplifies to the neat closed form 2^n. Finding this closed form requires a refined summation algorithm.

We call a sum *indefinite* if the upper summation bound is a variable that does not occur anywhere else in the summation problem. All other summation problems are called *definite*. In particular, a summation problem is called definite if there is a variable that occurs both in a summation bound and in the summand expression. So let us assume that we are given a bivariate hypergeometric sequence $(a_{n,k})_{n,k=0}^{\infty}$ in $\mathbb{K}$ and our goal is to simplify, if possible, the definite sum

$$s_n := \sum_{k=0}^{n} a_{n,k}.$$

Assume that $(s_n)_{n=0}^{\infty}$ is hypergeometric so that there are polynomials $c_0(t), c_1(t) \in \mathbb{K}[t]$ with

$$c_0(n)s_n + c_1(n)s_{n+1} = 0 \quad (n \geq 0).$$

For these we have

$$\sum_{k=0}^{n}(c_0(n)a_{n,k} + c_1(n)a_{n+1,k}) = -c_1(n)a_{n+1,n+1} \quad (n \geq 0).$$

Since $(a_{n,k})_{n=0}^{\infty}$ is assumed to be hypergeometric in both n and k, there are rational functions $u(t,x), v(t,x) \in \mathbb{K}(t,x)$ with $a_{n,k+1} = u(n,k)a_{n,k}$ and $a_{n+1,k} = v(n,k)a_{n,k}$ for all $n,k \in \mathbb{N}$ where $u(n,k)$ and $v(n,k)$ are defined. Therefore, for any $c_0(t), c_1(t) \in \mathbb{K}[t]$, we have

$$\bar{a}_{n,k} := c_0(n)a_{n,k} + c_1(n)a_{n+1,k} = (c_0(n) + c_1(n)v(n,k))a_{n,k}$$

for all $n,k \in \mathbb{N}$ where $v(n,k)$ is defined, and so $(\bar{a}_{n,k})_{n,k=0}^{\infty}$ is a bivariate hypergeometric sequence.

Zeilberger's idea is to invoke Gosper's algorithm on the sequence $(\bar{a}_{n,k})_{n,k=0}^{\infty}$ with

$$\bar{a}_{n,k} = c_0 a_{n,k} + c_1 a_{n+1,k},$$

viewing c_0, c_1 as additional variables and keeping n fixed. Of course we cannot expect that Gosper's algorithm applied to $(\bar{a}_{n,k})_{n,k=0}^{\infty}$ will provide us with some closed form that is valid for any choice of c_0, c_1. Instead, it will most often just tell us "no hypergeometric solution". The crucial observation is that Gosper's algorithm can be refined such as to determine during the computation how c_0 and c_1 can be chosen so that a closed form exists.

For $\bar{u}(t,x) := \frac{c_0 + c_1 v(t,x+1)}{c_0 + c_1 v(t,x)} u(t,x) \in \mathbb{K}(c_0, c_1, t)(x)$ we have $\bar{a}_{n,k+1} = \bar{u}(n,k)\bar{a}_{n,k}$ for all $n, k \in \mathbb{N}$ where $\bar{u}(n,k)$ is defined; this $\bar{u}(t,x)$ is the input to Gosper's algorithm. We next have to determine polynomials $p(t,x)$, $q(t,x)$, $r(t,x) \in \mathbb{K}(c_0, c_1, t)[x]$ such that

$$\bar{u}(t,x) = \frac{p(t,x+1)}{p(t,x)} \frac{q(t,x)}{r(t,x+1)}$$

and

$$\gcd_x(q(t,x), r(t,x+i)) = 1 \qquad (i \in \mathbb{N} \setminus \{0\}).$$

It is easy to see that we will have $p(t,x) = c_0 p_0(t,x) + c_1 p_1(t,x)$ for some polynomials $p_0(t,x), p_1(t,x) \in \mathbb{K}(t)[x]$ and that $q(t,x)$ and $r(t,x)$ will be free of c_0, c_1. Therefore the Gosper equation has the form

$$c_0 p_0(t,x) + c_1 p_1(t,x) = q(t,x)y(t,x+1) - r(t,x)y(t,x).$$

This equation is next to be solved simultaneously for a polynomial $y(t,x) \in \mathbb{K}(t)[x]$ and rational functions $c_0, c_1 \in \mathbb{K}(t)$. Recall now that the algorithm from Sect. 3.4 proceeds by making an ansatz

$$y(t,x) = y_0(t) + y_1(t)x + \cdots + y_d(t)x^d$$

for suitably chosen $d \in \mathbb{N}$, plugs this ansatz into the equation, compares coefficients with respect to x, and solves the resulting linear system. In our present situation, we can solve simultaneously for $y_0, \ldots, y_d$ and the two as yet undetermined coefficients c_0, c_1. This will give us a homogeneous linear system of equations with coefficients in $\mathbb{K}(t)$. Its solutions correspond to all the triples $(c_0(t), c_1(t), y(t,x)) \in \mathbb{K}(t)^2 \times \mathbb{K}(t)[x]$ satisfying our equation. This is the key point: allowing additional freedom on the left hand side in form of the additional variables c_0, c_1 can make a polynomial solution possible even when none would exist for a fixed left hand side.

Suppose $(c_0(t), c_1(t), y(t,x))$ is a solution of the Gosper equation where not both of $c_0(t), c_1(t)$ are zero. Then for the rational function $w(t,x) := \frac{r(t,x)}{p(t,x)} y(t,x)(c_0(t) + c_1(t)v(t,x)) \in \mathbb{K}(t,x)$ we have, by construction,

$$c_0(t) + c_1(t)v(t,x) = w(t,x+1)u(t,x) - w(t,x).$$

If this equation happens to remain meaningful when we substitute for t and x any specific integers $n, k \in \mathbb{N}$, or if it can be made meaningful by employing continuity arguments or other means, then this implies

$$c_0(n)a_{n,k} + c_1(n)a_{n+1,k} = w(n, k+1)a_{n,k+1} - w(n,k)a_{n,k} \qquad (n, k \in \mathbb{N}).$$

Summing this equation for k from 0 to m yields

$$\sum_{k=0}^{m} (c_0(n)a_{n,k} + c_1(n)a_{n+1,k}) = w(n, m+1)a_{n,m+1} - w(n,0)a_{n,0} \qquad (n, m \in \mathbb{N}).$$

Setting now m to n and adding $c_1(n)a_{n+1,n+1}$ to both sides gives

$$c_0(n)s_n + c_1(n)s_{n+1} = \text{rhs}_n \qquad (n \in \mathbb{N}),$$

where rhs_n refers to the resulting explicit linear combination of hypergeometric sequences. So at this point we have found an inhomogeneous recurrence equation for the definite sum s_n, and finding a closed form for the sum s_n amounts to finding a hypergeometric solution of this recurrence. (Gosper's algorithm can be used for this again, see Problem 5.10.) Even better: in many examples the right hand side rhs_n is identically zero, and there is actually no need to solve anything. The solution in this case is a hypergeometric sequence with shift quotient $-c_0(x)/c_1(x)$.

As an example, let us see how to rediscover Vandermonde's identity. For formal parameters λ and μ, let $a_{n,k} := \binom{\lambda}{k}\binom{\mu}{n-k} \in \mathbb{Q}(\lambda, \mu)$. We want to simplify the definite sum

$$s_n := \sum_{k=0}^{n} a_{n,k} \qquad (n \geq 0).$$

Following Zeilberger's advice, we will look for c_0, c_1 such that for $\bar{a}_{n,k} := c_0 a_{n,k} + c_1 a_{n+1,k}$ the indefinite sum

$$\sum_{k=0}^{m} \bar{a}_{n,k}$$

admits a hypergeometric closed form. We have $\bar{a}_{n,k+1} = \bar{u}(n,k)\bar{a}_{n,k}$ for

$$\bar{u}(t,x) = \frac{\overbrace{\left((x-t)c_0 + (-x-1-\mu+t)c_1\right)}^{=p(t,x+1)}\overbrace{(x-\lambda)(x-t-1)}^{=q(x,t)}}{\underbrace{\left((x-1-t)c_0 + (-x-\mu+t)c_1\right)}_{=p(t,x)}\underbrace{(x+1)(x+1+\mu-t)}_{=r(x,t+1)}} \in \mathbb{Q}(\lambda,\mu)(t,x)$$

and all $n, k \in \mathbb{N}$ where $\bar{u}(n,k)$ is defined. The rational function $\bar{u}(t,x)$ decomposes into a Gosper form as indicated. Next we have to find $y(x) \in \mathbb{Q}(\lambda, \mu, t)[x]$ satisfying

$$(x-1-t)c_0 + (-x-\mu+t)c_1 = (x-\lambda)(x-n-1)y(x+1) - x(x+\mu-t)y(x).$$

From the third case of the degree analysis in Sect. 3.4 we find that any polynomial solution must have degree $d = \deg p(x) - \deg q(x) + 1 = 1 - 2 + 1 = 0$. Making an ansatz $y(x) = y_0$ for some undetermined $y_0 \in \mathbb{Q}(\lambda, \mu)(t)$ and comparing coefficients with respect to x we are led to the linear system

$$\begin{pmatrix} -t-1 & t-\mu & -\lambda t - \lambda \\ 1 & -1 & 1 + \lambda + \mu \end{pmatrix} \begin{pmatrix} c_0 \\ c_1 \\ y_0 \end{pmatrix} = 0.$$

The solution vector $(c_0, c_1, y_0) = (t - \lambda - \mu, t + 1, 1)$ yields, after some simplification, the relation

$$(n - \lambda - \mu)a_{n,k} + (n+1)a_{n+1,k} = \frac{(k+1)(n-k-1-\mu)}{n-k}a_{n,k+1} - \frac{k(n-k-\mu)}{n-k+1}a_{n,k}$$

which we would next like to sum for k from 0 to n. Unfortunately, the right hand side is undefined for $k = n$, so we must be careful. We could argue that by regarding n and k as complex variables and λ, μ as fixed complex numbers, the singularities at $n = k$ are removable because the limiting value for $n \to k$ of the first term on the right hand side is zero. In the understanding that a continuous extension has been done accordingly, we can safely take the sum for k from 0 to n, obtaining, after some further simplification,

$$\sum_{k=0}^{n} \left((n - \lambda - \mu)a_{n,k} + (n+1)a_{n+1,k} \right) = (n - \lambda)\binom{\lambda}{n} \qquad (n \geq 0).$$

Adding $(n+1)a_{n+1,n+1}$ to both sides doing some more simplification, we arrive at

$$(n - \lambda - \mu)s_n + (n+1)s_{n+1} = 0 \qquad (n \geq 0).$$

This recurrence has the solution $\binom{\lambda+\mu}{n}$, and since this term agrees with the sum for $n = 0$, we obtain that

$$\sum_{k=0}^{n} \binom{\lambda}{k}\binom{\mu}{n-k} = \binom{\lambda+\mu}{n} \qquad (n \geq 0),$$

as expected.

Our discussion so far implies only that a recurrence found by the algorithm is correct, and we have seen in the example that it found what we expected to find for Vandermonde's identity. But can we be sure that the algorithm will always discover a recurrence? Not quite. There are definite hypergeometric sums which simply do not satisfy any first order recurrence, and for such sums, our algorithm should better not find any. In such cases, the algorithm will tell us that the only possible choice for c_0 and c_1 is to set both to zero. And, worse, the algorithm may even overlook recurrence equations and tell us to set $c_0 = c_1 = 0$ even when a nontrivial recurrence for the sum under consideration exists. Fortunately, there is a simple extension that makes the algorithm complete.

Theorem 5.1 *Let $(a_{n,k})_{n,k=0}^{\infty}$ be a proper hypergeometric sequence in $\mathbb{C}$. Then there exist $r \in \mathbb{N}$ and $c_0(t), \ldots, c_r(t) \in \mathbb{C}[t]$, not all zero, and a rational function $w(t,x) \in \mathbb{C}(t,x)$ such that*

$$c_0(n)a_{n,k} + c_1(n)a_{n+1,k} + \cdots + c_r(n)a_{n+r,k} = w(n,k+1)a_{n,k+1} - w(n,k)a_{n,k}$$

for all $n,k \in \mathbb{N}$ where $w(n,k)$ and $w(n,k+1)$ are defined.

In a nutshell, the theorem says that every definite sum of a proper hypergeometric sequence satisfies a linear recurrence of some order r with rational function coefficients, and Zeilberger's algorithm is bound to discover one, provided that it is applied with $c_0, c_1, \ldots, c_r$ (r large enough) instead of just c_0 and c_1. For a proof we refer to the book A=B by Petkovšek, Wilf and Zeilberger [44].

5.6 Applications

The Hypergeometric Probability Distribution

Products of binomials frequently arise in probability theory. For example, consider an urn containing N balls, m green ones and $N - m$ blue ones. If we select a ball at random, we will clearly hit a green ball with probability m/N and a blue ball with probability $(N - m)/N$. If we select two balls at random, the probability that both of them are green is $m(m - 1)/(N(N - 1))$, and the probability that both of them are blue is $(N - m)(N - m - 1)/(N(N - 1))$. The probability of hitting one green and one blue ball is $2m(N - m)/(N(N - 1))$. In general, if we select n balls, and let X be the random variable that counts the number of green balls in the selected sample, then we have

$$P(X = k) = \frac{\binom{m}{k}\binom{N-m}{n-k}}{\binom{N}{n}}.$$

This is called the *hypergeometric probability distribution*. Analyzing its features amounts to solving some hypergeometric summation exercises.

First, to see that the probabilies sum up to 1, as they should, observe that

$$\sum_{k=0}^{n} P(X = k) = \sum_{k=0}^{n} \frac{\binom{m}{k}\binom{N-m}{n-k}}{\binom{N}{n}} = 1$$

is just a reformulation of Vandermonde's identity.

Second, for the *mean* of the distribution we obtain

$$E(X) := \sum_{k=0}^{n} kP(X = k) = \sum_{k=0}^{n} k\frac{\binom{m}{k}\binom{N-m}{n-k}}{\binom{N}{n}} = \frac{nm}{N},$$

where Zeilberger's algorithm will do the last step for us.

Third, for the *variance* we find

$$\mathrm{Var}(X) := \mathrm{E}(X^2) - \mathrm{E}(X)^2 = \sum_{k=0}^{n} \left(k - \frac{nm}{N}\right)^2 \frac{\binom{m}{k}\binom{N-m}{n-k}}{\binom{N}{n}} = \frac{mn(N-n)(N-m)}{N^2(N-1)},$$

and again Zeilberger's algorithm saves us from having to perform any hand calculations.

Elliptic Arc Length

For fixed $a, b > 0$, consider the ellipsis $E = \{(a\cos(\phi), b\sin(\phi)) : \phi \in [0, 2\pi]\}$. We are interested in the circumference $L(a, b)$ of E. For a circle, we know $L(a, a) = 2\pi a$, but if $a \neq b$, there is no closed form for $L(a, b)$ in terms of elementary functions. Still, it is possible to express $L(a, b)$ in terms of a hypergeometric series. Starting from the standard integral formula for arc length, we have

$$L(a, b) = \int_0^{2\pi} \sqrt{a^2 \sin(\phi)^2 + b^2 \cos(\phi)^2}\, d\phi \qquad \text{(by definition)}$$

$$= 4a \int_0^{\pi/2} \sqrt{1 + \left(\frac{b^2}{a^2} - 1\right)\cos(\phi)^2}\, d\phi \qquad \text{(by symmetry)}$$

$$= 4a \int_0^{\pi/2} \sum_{n=0}^{\infty} \binom{1/2}{n} \left(\frac{b^2}{a^2} - 1\right)^n \cos(\phi)^{2n}\, d\phi \qquad \text{(by the binomial theorem)}$$

$$= 4a \sum_{n=0}^{\infty} \binom{1/2}{n} \left(\frac{b^2}{a^2} - 1\right)^n \int_0^{\pi/2} \cos(\phi)^{2n}\, d\phi \qquad \begin{array}{l}\text{(by flipping sum}\\\text{and integral)}\end{array}$$

$$= 4a \sum_{n=0}^{\infty} \binom{1/2}{n} \left(\frac{b^2}{a^2} - 1\right)^n \frac{\pi}{2} \frac{\left(\frac{1}{2}\right)^{\overline{n}}}{n!} \qquad \text{(by Problem 5.16)}$$

$$= 2\pi a\, {}_2F_1\left(\begin{array}{c}-\frac{1}{2}, \frac{1}{2}\\1\end{array}\middle|\, 1 - \frac{b^2}{a^2}\right) \qquad \text{(by pattern matching)}$$

$$= \pi(a+b)\, {}_2F_1\left(\begin{array}{c}-\frac{1}{2}, -\frac{1}{2}\\1\end{array}\middle|\, \frac{(a-b)^2}{(a+b)^2}\right) \qquad \text{(by Landen's transform)}.$$

When a and b are close to each other, then this hypergeometric series converges quickly, and we can expect good approximations for $L(a, b)$ from functions whose Taylor expansion at the origin agrees with the above series to many terms. For example, with $\lambda = \frac{a-b}{a+b}$, Ramanujan pointed out that

$$\pi(a+b)\, {}_2F_1\left(\begin{array}{c}-\frac{1}{2}, -\frac{1}{2}\\1\end{array}\middle|\, \lambda^2\right) = 1 + \frac{3\lambda^2}{10 + \sqrt{4 - 3\lambda^2}} + O(\lambda^{10}) \qquad (\lambda \to 0),$$

an approximation which is both simple and accurate.

Monthly Problems

There are scientific journals with regular sections where readers can pose problems to challenge other readers. Binomial sums originating from all corners of the mathematical universe are a frequent topic in such problem sections. Some of the problems can be solved by directly applying the techniques of this chapter. And even when the techniques do not apply directly, they are typically still helpful somewhere during the solving process. Let us solve two example problems that have appeared in 1995 in the problem section of the *American Mathematical Monthly*.

The first problem was posed as number 10424 by Ira Gessel: *Evaluate the sum*

$$s_n := \sum_{0 \le k \le n/3} \frac{n}{n-k} 2^k \binom{n-k}{2k} \quad (n \ge 1).$$

To solve this problem, first use Zeilberger's algorithm to determine a recurrence for $(s_n)_{n=0}^{\infty}$. This gives

$$s_{n+3} - 2s_{n+2} + s_{n+1} - 2s_n = 0 \quad (n \ge 1).$$

This recurrence accidentally happens to be C-finite and can therefore be solved in closed form as described in the previous chapter. The solution is $s_n = \frac{1}{2}(2^n + i^n + (-i)^n)$ where $i = \sqrt{-1}$.

The second problem was posed as number 10473 by Emre Alkan: *Prove that there are infinitely many positive integers n such that*

$$s_n := \sum_{k=0}^{n} \frac{1}{5} \binom{2n+1}{2k} 3^k 2^{-n}$$

is an odd integer. We start again by applying Zeilberger's algorithm. Again it happens to find a C-finite recurrence. This time the recurrence reads

$$s_{n+2} - 4s_{n+1} + s_n = 0 \quad (n \ge 1).$$

The sequence $(s_n)_{n=0}^{\infty}$ starts like

$$\tfrac{1}{5}, 1, \tfrac{19}{5}, \tfrac{71}{5}, 53, \tfrac{989}{5}, \tfrac{3691}{5}, 2755, \tfrac{51409}{5}, \tfrac{191861}{5}, 143207, \tfrac{2672279}{5}, \tfrac{9973081}{5}, 7444009, \ldots$$

which suggests the conjecture that every $(3n+1)$-th term might be an odd integer. Using the C-finite recurrence for $(s_n)_{n=0}^{\infty}$ as input, the algorithm behind Theorem 4.2 tells us that $(a_n)_{n=0}^{\infty}$ with $a_n := s_{3n+1}$ satisfies

$$a_{n+2} - 52a_{n+1} + a_n = 0 \quad (n \ge 1).$$

Therefore, if a_n and a_{n+1} are odd integers, then so is a_{n+2}. Since $a_0 = s_1 = 1$ and $a_1 = s_4 = 53$ are both odd, it follows by induction that all the a_n are odd integers, as conjectured.

5.7 Problems

Problem 5.1 Let $f\colon \mathbb{N}^2 \to \mathbb{Z}$ be any function. Show that the following properties are equivalent:

1. $f(n,k) = \binom{n}{k}$ for all $n,k \in \mathbb{N}$ (with $\binom{n}{k}$ as defined in the text).
2. $f(n,k) = f(n-1,k) + f(n-1,k-1)$ for all $n,k > 0$ and $f(n,0) = 1$ $(n \geq 0)$, $f(0,n) = 0$ $(n > 0)$.
3. For all $n,k \in \mathbb{N}$, the number of subsets of $\{1,2,\dots,n\}$ with exactly k elements is $f(n,k)$.
4. $(a+b)^n = \sum_{k=0}^{n} f(n,k) a^k b^{n-k}$ for all $n,k \in \mathbb{N}$ and $a,b \in \mathbb{K}$.

Problem 5.2 It is clear that when $p(x) \in \mathbb{K}[x]$ is a polynomial with $p(n) = 0$ for all $n \in \mathbb{N}$, then $p(x)$ must be the zero polynomial, because for a nonzero polynomial, the number of zeros is bounded by its (finite) degree. (Recall that we assume throughout that $\mathbb{K}$ has characteristic zero.) A bivariate polynomial $p(x,y)$, however, may have infinitely many zeros. (Think of $p(x,y) = x - y$.) Still: prove that if $p(n,m) = 0$ for all $n,m \in \mathbb{N}$, then $p(x,y)$ must be the zero polynomial.

Problem 5.3 Determine all sequences which are at the same time hypergeometric and C-finite.

Problem 5.4 Determine the asymptotics of

$$\frac{2^n}{n+1} \binom{2n}{n}^2 \binom{5n}{2n}^{-1} \binom{3n}{n+1}.$$

Problem 5.5 In $\mathbb{Q}(a,b,c)[[x]]$, show that $f(x) = {}_2F_1\left(\begin{matrix} a,b \\ c \end{matrix}\middle| x\right)$ satisfies the differential equation

$$x(1-x)D_x^2 f(x) + (c - (a+b+1)x)D_x f(x) - a\,bf(x) = 0.$$

Problem 5.6 1. In $\mathbb{Q}(a,b,c)[[x]]$, prove Pfaff's hypergeometric transformation

$${}_2F_1\left(\begin{matrix} a,b \\ c \end{matrix}\middle| x\right) = (1-x)^{-a}\,{}_2F_1\left(\begin{matrix} a,c-b \\ c \end{matrix}\middle| \frac{x}{x-1}\right).$$

(*Hint:* First multiply both sides by $(1-x)^a$ and then derive sum representations for the coefficient of x^n on both sides. Then use Zeilberger's algorithm to prove the resulting summation identity.)

2. In $\mathbb{Q}(a,b,c)[[x]]$, prove Euler's hypergeometric transformation

$${}_2F_1\left(\begin{matrix} a,b \\ c \end{matrix}\middle| x\right) = (1-x)^{c-a-b}\,{}_2F_1\left(\begin{matrix} c-a,c-b \\ c \end{matrix}\middle| x\right)$$

by applying Pfaff's transformation twice.

3. Derive the Pfaff-Saalschütz identity from Euler's transformation.

Problem 5.7 In $\mathbb{Q}(a,b,c)[[x]]$, prove the relation

$$_2F_1\left(\begin{matrix} a,b \\ c \end{matrix}\middle| x\right) = _2F_1\left(\begin{matrix} a+1,b \\ c \end{matrix}\middle| x\right) - \frac{bx}{c}\,_2F_1\left(\begin{matrix} a+1,b+1 \\ c+1 \end{matrix}\middle| x\right).$$

Problem 5.8 Let $u_1(x), u_2(x) \in \mathbb{K}[x]$ with $\gcd(u_1(x), u_2(x)) = 1$.

1. Show that there are at most finitely many $i \in \mathbb{N} \setminus \{0\}$ with $\gcd(u_1(x), u_2(x+i)) \neq 1$.

2. Let $i \in \mathbb{N} \setminus \{0\}$ be the greatest index such that $\gcd(u_1(x), u_2(x+i)) \neq 1$, and let

$$g(x) = \gcd(u_1(x), u_2(x+i)).$$

Set $\bar{u}_1(x) = u_1(x)/g(x)$, $\bar{u}_2(x) = u_2(x)/g(x-i)$ and $p(x) = g(x-1)\cdots g(x-i)$. Show that

$$\frac{u_1(x)}{u_2(x)} = \frac{p(x+1)}{p(x)}\frac{\bar{u}_1(x)}{\bar{u}_2(x)},$$

and that if $j \in \mathbb{N}$ is such that $\gcd(\bar{u}_1(x), \bar{u}_2(x+j)) \neq 1$ then $j < i$.

3. Assume that there is a way to find integer roots of univariate polynomials over $\mathbb{K}$. In the next chapter (p. 124) we will introduce the *resultant* $\operatorname{res}_x(p(x), q(x))$ of two polynomials $p(x), q(x) \in \mathbb{K}[x]$. Its key features are that $\gcd(p(x), q(x)) = 1 \iff \operatorname{res}_x(p(x), q(x)) \neq 0$ for any $p(x), q(x) \in \mathbb{K}[x]$ and that $\operatorname{res}_x(p(x), q(x+t))$ is a polynomial in t only. Use this to show that one can determine, given $u_1(x), u_2(x) \in \mathbb{K}[x]$, all indices $i \in \mathbb{N} \setminus \{0\}$ with $\gcd(u_1(x), u_2(x+i)) \neq 1$.

4. Give an algorithm that computes a Gosper form of a given rational function $u(x)$ by repeatedly applying the previous two observations.

5. Implement your algorithm in a computer algebra system of your choice.

Problem 5.9 Can it happen that the Gosper equation has more than one solution? And if so, how does the choice of the polynomial solution $y(x)$ in the algorithm influence the final output?

Problem 5.10 Design an algorithm which takes as input two rational functions $c_0(x), c_1(x) \in \mathbb{K}(x)$ and a hypergeometric sequence $(a_n)_{n=0}^\infty$ and which decides whether the equation

$$c_1(n)s_{n+1} + c_0(n)s_n = a_n$$

has a hypergeometric solution $(s_n)_{n=0}^\infty$. Assume for simplicity that $c_1(n) \neq 0 \neq c_0(n)$ for all $n \in \mathbb{N}$.

(*Hint:* Think of a substitution that turns the equation into a telescoping equation and then apply Gosper's algorithm.)

Problem 5.11 Evaluate the following sums:

1. $\displaystyle\sum_{k=0}^{n} \binom{x+k}{k}$

2. $\displaystyle\sum_{k=0}^{n} \binom{2k}{k} 4^{-k}$

3. $\displaystyle\sum_{k=0}^{n} \frac{1}{k^2 + \sqrt{5}k - 1}$

4. $\displaystyle\sum_{k=0}^{n} (4k+1) \frac{k!}{(2k+1)!}$

5. $\displaystyle\sum_{k=0}^{n} \frac{k^2}{(k+1)(k+2)} 4^k$

6. $\displaystyle\sum_{k=0}^{n} \frac{(4k-1)}{(2k-1)^2} 16^{-k} \binom{2k}{k}^2$

Problem 5.12 Use Gosper's algorithm to prove that $(H_n)_{n=0}^{\infty}$ is not hypergeometric.

Problem 5.13 Find a nonzero polynomial $p(x) \in \mathbb{K}[x]$ of minimal degree such that

$$\sum_{k=0}^{n} \frac{p(k)}{k!}$$

has a hypergeometric closed form.

Problem 5.14 Evaluate the following sums

1. $\displaystyle\sum_{k=0}^{n} k \binom{2n+1}{2k+1}$

2. $\displaystyle\sum_{k=0}^{n} \binom{n+k}{2k} (-4)^k$

3. $\displaystyle\sum_{k=0}^{n} \binom{n}{2k} \binom{2k}{k} 4^{-k}$

4. $\displaystyle\sum_{k=0}^{n} (-1)^k \binom{n}{k} \binom{2n-2k}{n}$

5. $\displaystyle\sum_{k=0}^{2n} \binom{2n}{k} \binom{2k}{k} (-\tfrac{1}{2})^k$

6. $\displaystyle\sum_{k=0}^{n} \binom{n+k}{2k} \binom{2k}{k} \frac{(-1)^k}{k+1}$

Problem 5.15 The sum

$$\sum_{k=0}^{n} \binom{n}{k}^2 \binom{n+k}{k}^2$$

satisfies a second order recurrence. Find it.

Problem 5.16 Prove Wallis integral identity

$$\int_0^{\pi/2} \cos(\phi)^{2n} \, d\phi = \frac{\pi}{2} \frac{(\tfrac{1}{2})^{\overline{n}}}{n!} \qquad (n \geq 0).$$

(*Hint:* First find polynomials $c_0(t), c_1(t) \in \mathbb{Q}[t]$ such that the indefinite integral $\int (c_0(n) \cos(\phi)^{2n} + c_1(n) \cos(\phi)^{2n+2}) \, d\phi$ can be evaluated in closed form.)

Chapter 6
Algebraic Functions

The next station in our tour through various classes of formal power series is the class of algebraic power series. These power series share the property that they can be described as solutions of a polynomial equation. A deep theory has been developed for these power series mainly because of the important role they play in the field of algebraic geometry for locally describing algebraic curves and surfaces in the neighborhood of a specific point. But the significance of algebraic power series is by no means restricted to algebraic geometry. In combinatorics, for example, they turn out to play a prominent role as well, as they do appear as generating functions for a number of important combinatorial objects.

6.1 Catalan Numbers

Now is the time to return to the question of how many plane binary trees with n internal nodes there are. We have considered this question already in Sect. 2.6. We have learned that these numbers are called the Catalan numbers and commonly denoted by C_n, we have seen that their first terms are

$$1, \ 1, \ 2, \ 5, \ 14, \ 42, \ 132, \ 429, \ 1430, \ 4862, \ 16796, \ldots$$

and we have observed that there is some regularity among these numbers. Specifically, using automated guessing, we have found that the generating function $C(x) = \sum_{n=0}^{\infty} C_n x^n$ appears to satisfy the equation

$$1 - C(x) + xC(x)^2 = 0.$$

How can we prove that this guess is correct? Well, what do we know about the C_n after all? They are defined as the numbers of certain trees, and given that trees are recursively defined objects, it is not too unreasonable to hope for a recurrence relation for the C_n. To wit, a plane binary tree is either a single node (recursion base) or a composition of a root node and two other plane binary trees (recursion step).

M. Kauers, P. Paule, *The Concrete Tetrahedron*
© Springer-Verlag/Wien 2011

A tree with n internal nodes will consist of a root node, one subtree with k nodes, and one subtree with $(n-1)-k$ nodes, for some k. This combinatorial consideration translates immediately into the recurrence

$$C_n = \sum_{k=0}^{n-1} C_k C_{(n-1)-k} \qquad (n \geq 1)$$

for the number of trees. Going from here, as we have often done already, to the generating function, we obtain (taking into account $C_0 = 1$)

$$C(x) = 1 + \sum_{n=1}^{\infty} \left(\sum_{k=0}^{n-1} C_k C_{(n-1)-k} \right) x^n = 1 + xC(x)^2,$$

as expected. This confirms that (as usual) the automated guess was correct. Solving the equation for $C(x)$, we get the closed form representation

$$C(x) = \frac{1 - \sqrt{1-4x}}{2x}.$$

Also the Catalan numbers C_n themselves have a closed form. An automated guesser will have no difficulties in conjecturing the recurrence relation

$$(n+2)C_{n+1} - 2(2n+1)C_n = 0 \qquad (n \geq 0),$$

which suggests $C_n = \frac{1}{n+1}\binom{2n}{n}$. To formally verify this conjecture, all we need to do is to check compatibility with the recurrence we derived before (and with the value at $n = 0$). This means we have to prove the identity

$$\sum_{k=0}^{n-1} \frac{1}{k+1}\binom{2k}{k} \frac{1}{(n-1)-k+1} \binom{2(n-1-k)}{n-1-k} = \frac{1}{n+1}\binom{2n}{n} \qquad (n \geq 1),$$

which of course we can do without wasting a single thought on it by appealing to Zeilberger's algorithm. (Incidentally, we could have deduced the closed form for C_n also from the closed form of $C(x)$ by using the binomial theorem and doing some straightforward simplifications, but why should we go through a boring, time-consuming, and error-prone hand calculation if we don't have to?)

Let us now turn from plane binary trees to some other combinatorial objects, known as Dyck paths. These are paths in a two-dimensional lattice, starting in the origin, ending somewhere on the horizontal axis, consisting only of steps either right-up or right-down, and never stepping below the horizontal axis. Two sample Dyck paths are shown in Fig. 6.1.

How many Dyck paths are there that end at the point $(n, 0)$? There obviously cannot be any if n is odd. Otherwise, for $n = 0$, there is one, the empty path. For $n = 2$, there is one as well, as the only possible path consists of a step right-up followed by a step right-down. For $n = 4$, we have no choice for the first step (it must be right-up), but we have a choice for the second step (right-up or right-down), and each choice

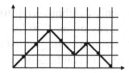 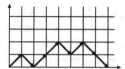

Fig. 6.1 Two Dyck paths with 10 steps

determines uniquely the rest of the path, so there are two Dyck paths for $n = 4$. For the general solution, it is convenient to start from the more general number $D_{n,m}$ of all Dyck-like paths that end at point (n,m). Clearly, we have $D_{n,m} = 0$ if $n < 0$ or $m < 0$ and $D_{0,0} = 1$ by the combinatorial definition. For $(n,m) \neq (0,0)$ the combinatorial definition also directly explains the recurrence equation

$$D_{n,m} = D_{n-1,m-1} + D_{n-1,m+1}.$$

If we let $D(x,y) := \sum_{n,m=0}^{\infty} D_{n,m} x^n y^m$ be the bivariate generating function, then this recurrence equation implies

$$\sum_{n,m=0}^{\infty} D_{n,m} x^n y^m = 1 + \sum_{n,m=0}^{\infty} D_{n-1,m-1} x^n y^m + \sum_{n,m=0}^{\infty} D_{n-1,m+1} x^n y^m$$

$$= 1 + xy \sum_{n,m=0}^{\infty} D_{n-1,m-1} x^{n-1} y^{m-1} + \frac{x}{y} \sum_{n,m=0}^{\infty} D_{n-1,m+1} x^{n-1} y^{m+1}$$

$$= 1 + xy \underbrace{\sum_{\substack{n=-1 \\ m=-1}}^{\infty} D_{n,m} x^n y^m}_{=D(x,y)} + \frac{x}{y} \underbrace{\sum_{\substack{n=-1 \\ m=1}}^{\infty} D_{n,m} x^n y^m}_{=D(x,y)-D(x,0)},$$

and finally

$$(y - x - xy^2) D(x,y) = y - xD(x,0).$$

And now comes a trick! In order to solve this functional equation, we tie up x and y in such a way that the left hand side vanishes. This is easily done by setting $y = (1 - \sqrt{1 - 4x^2})/(2x)$. Note that this is a power series in x of positive order. This substitution leads us directly to

$$D(x,0) = y/x = \frac{1 - \sqrt{1 - 4x^2}}{2x^2} = C(x^2).$$

From this we can conclude that the number of Dyck paths of length $2n$ is exactly the n-th Catalan number C_n. In addition, we get for free that the number of Dyck-like paths ending at level $m \geq 0$ is given by the coefficients of the bivariate power series

$$D(x,y) = \frac{y - xC(x^2)}{y - x - xy^2}.$$

Dyck paths and plane binary trees are not the only combinatorial structures which are counted by Catalan numbers, but there are many more. A collection of some 66 of them can be found in Exercise 6.19 of [54]. It is not exaggerated to say that the Catalan numbers are the most prominent sequence in combinatorics.

6.2 Basic Facts and Definitions

Catalan numbers belong to the class of sequences whose generating function is algebraic. In general, $a(x) \in \mathbb{K}[[x]]$ is called *algebraic* if there exist polynomials $p_0(x), \ldots, p_d(x) \in \mathbb{K}[x] \subseteq \mathbb{K}[[x]]$, not all zero, such that

$$p_0(x) + p_1(x)a(x) + p_2(x)a(x)^2 + \cdots + p_d(x)a(x)^d = 0.$$

This is the case, as we have seen, for the generating function of the Catalan numbers. It does not make much of a difference whether we request the $p_i(x)$ to be polynomials or rational functions. Taking them as rational function has the advantage that we can regard

$$p(x,y) := p_0(x) + p_1(x)y + \cdots + p_d(x)y^d$$

as a univariate polynomial in y over the coefficient field $\mathbb{K}(x)$. A substitution $p(x,a(x))$ for $a(x) \in \mathbb{K}[[x]]$ is then meant to take place in the bigger domain $\mathbb{K}((x))$ which contains both $\mathbb{K}(x)$ and $\mathbb{K}[[x]]$ as subrings. This alternative point of view can sometimes be more handy than working with bivariate polynomials over $\mathbb{K}$. For example, we can exploit facts about greatest common divisors of univariate polynomials in order to understand how the various algebraic equations satisfied by a fixed algebraic power series are related to each other, and we can use the Euclidean algorithm for doing actual computations. For a fixed algebraic power series $a(x) \in \mathbb{K}[[x]]$, let $A \subseteq \mathbb{K}(x)[y]$ be the set of all *annihilating polynomials*, i.e., all polynomials $p(x,y)$ such that $p(x,a(x)) = 0$. Then we obviously have

$$p(x,a(x)) + q(x,a(x)) = 0 \quad \text{and} \quad r(x,a(x))p(x,a(x)) = 0$$

for all polynomials $p(x,y), q(x,y) \in A$ and all $r(x,y) \in \mathbb{K}(x)[y]$. This means that A is an *ideal* in $\mathbb{K}(x)[y]$, and this in turn means that there exists some polynomial $m(x,y) \in \mathbb{K}(x)[y]$ such that every polynomial in A is a multiple of $m(x,y)$. If we further impose that $m(x,y)$ be monic (i.e., that the leading coefficient of $m(x,y)$ with respect to y should be 1), then $m(x,y)$ is uniquely determined. This unique polynomial $m(x,y)$ is called the *minimal polynomial* of $a(x)$.

The minimal polynomial must necessarily be an irreducible polynomial, for if we could split it nontrivially into two factors,

$$m(x,y) = u(x,y)v(x,y),$$

then we must have $u(x,a(x)) = 0$ or $v(x,a(x)) = 0$, because $\mathbb{K}((x))$ is a field, and so $u(x,y)$ or $v(x,y)$ would be an element of A although none of them can be a multiple of $m(x,y)$.

If $a(x)$ is an algebraic power series whose minimal polynomial $m(x,y)$ has degree d (with respect to y), then by polynomial division with remainder we can find for every polynomial $p(x,y) \in \mathbb{K}(x)[y]$ two other polynomials $q(x,y), r(x,y) \in \mathbb{K}(x)[y]$ with

$$p(x,y) = q(x,y)m(x,y) + r(x,y)$$

where the degree of $r(x,y)$ is less than d. Setting $y = a(x)$ gives $p(x,a(x)) = r(x,a(x))$. In other words, every polynomial expression $p(x,a(x))$ can be "reduced" to an equivalent polynomial expression where $a(x)$ appears only with degree less than d. In yet other words, the $\mathbb{K}(x)$-vector space

$$V := \mathbb{K}(x) \oplus a(x)\mathbb{K}(x) \oplus a(x)^2 \mathbb{K}(x) \oplus \cdots \oplus a(x)^{d-1} \mathbb{K}(x) \subseteq \mathbb{K}((x)),$$

where $\oplus$ is meant to denote the direct sum of vector spaces, is closed under multiplication, and therefore a commutative ring with 1.

And not only this. If $p(x,y) \in \mathbb{K}(x)[y]$ is nonzero with $\deg_y p(x,y) < d$, then, since $m(x,y)$ is irreducible, we have $\gcd_y(m(x,y), p(x,y)) = 1$. Therefore, by the extended Euclidean algorithm, we can find polynomials $u(x,y), v(x,y) \in \mathbb{K}(x)[y]$ with

$$u(x,y)m(x,y) + v(x,y)p(x,y) = 1,$$

from which by setting $y = a(x)$ we get $1/p(x,a(x)) = v(x,a(x))$. In other words, the reciprocal of an element of V can be expressed as a polynomial in $a(x)$ with rational function coefficients in x. In yet other words, V is closed under division, and therefore even a field.

And this is still not all. If we differentiate the equation

$$m(x,a(x)) = 0$$

with respect to x, we get

$$m_1(x,a(x)) + m_2(x,a(x))a'(x) = 0$$

where by $m_1(x,y)$ and $m_2(x,y)$ we denote the partial derivatives $D_x m(x,y), D_y m(x,y)$ of $m(x,y)$ with respect to x and y, respectively. Differentiation with respect to y will certainly decrease the degree in y, so that we surely have $\gcd_y(m(x,y), m_2(x,y)) = 1$, again because $m(x,y)$ is irreducible. Again we can use the extended Euclidean algorithm to find $u(x,y)$ and $v(x,y)$ with

$$u(x,y)m(x,y) + v(x,y)m_2(x,y) = 1$$

from which we obtain $a'(x) = -v(x,a(x))m_1(x,a(x))$. In other words, the derivative of an algebraic power series $a(x)$ can be expressed as a polynomial in $a(x)$ with rational function coefficients in x. In yet other words, V is closed under differentiation, and therefore a *differential field*.

Here is an important structural consequence of these observations.

Theorem 6.1 *Every algebraic power series in* $\mathbb{K}[[x]]$ *satisfies a linear differential equation with polynomial coefficients in* $\mathbb{K}[x]$.

Proof. Let $a(x)$ be an algebraic power series and let d be the degree of its minimal polynomial. Then, as observed before,

$$V := \mathbb{K}(x) \oplus a(x)\,\mathbb{K}(x) \oplus a(x)^2\,\mathbb{K}(x) \oplus \cdots \oplus a(x)^{d-1}\,\mathbb{K}(x)$$

is a differential field, so the power series

$$a(x), a'(x), a''(x), a'''(x), \ldots$$

all belong to V. At the same time, V is a $\mathbb{K}(x)$-vector space of dimension d. Any $d+1$ vectors in a vector space of dimension d must be linearly dependent, and so in particular $a(x), a'(x), \ldots, a^{(d)}(x)$ must be linearly dependent over $\mathbb{K}(x)$, which means nothing else but the existence of a nontrivial relation

$$p_0(x)a(x) + p_1(x)a'(x) + \cdots + p_d(x)a^{(d)}(x) = 0.$$

Multiplying the equation by the common denominator of the rational functions $p_i(x)$ gives the desired equation. $\qquad\square$

In what we have said so far, we have hardly had a need to take into account that we are talking about formal power series. In fact, the observations we made do not depend on this but can be formulated more generally: whenever $m(x,y) \in \mathbb{K}[x,y]$ is irreducible as element of $\mathbb{K}(x)[y]$, then the factor ring $R := \mathbb{K}(x)[y]/\langle m(x,y)\rangle$ together with the derivation $Dx := 1$ and $Dy := -D_x m(x,y)/D_y m(x,y)$ is a differential field. The ring R is called an *algebraic extension* of the field $\mathbb{K}(x)$ and its generator y is a purely algebraic object playing the role of an algebraic power series.

Still, we are primarily interested in actual power series. Can we associate to every irreducible polynomial $m(x,y) \in \mathbb{K}[x,y]$ a power series $a(x)$ with $m(x,a(x)) = 0$? And if so, is $a(x)$ uniquely determined by its minimal polynomial? The answer to both questions is "yes, but...". In almost all cases (including most cases arising in applications), a power series *is* uniquely determined by the polynomial equations it satisfies, and usually the implicit function theorem for formal power series (Theorem 2.9) applied to $m(x,y) \in \mathbb{K}[x,y] \subseteq \mathbb{K}[[x,y]]$ asserts this fact. *But* there are degenerate situations in which no power series solution exists, or several, or there is a unique solution though the implicit function theorem it is not applicable.

There is an intuitive geometric interpretation of such degenerate cases that becomes apparent if for the specific polynomial $m(x,y)$ at hand we look at the curve consisting of all points (x,y) with $m(x,y) = 0$. Power series $a(x)$ satisfying $m(x,a(x)) = 0$ and $a(0) = 0$ correspond to branches of that curve traversing the origin with a finite slope. Examples for some of the possible situations are given in Fig. 6.2.

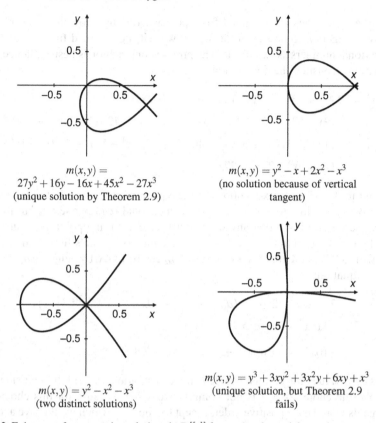

$$m(x,y) =$$
$$27y^2 + 16y - 16x + 45x^2 - 27x^3$$
(unique solution by Theorem 2.9)

$$m(x,y) = y^2 - x + 2x^2 - x^3$$
(no solution because of vertical tangent)

$$m(x,y) = y^2 - x^2 - x^3$$
(two distinct solutions)

$$m(x,y) = y^3 + 3xy^2 + 3x^2y + 6xy + x^3$$
(unique solution, but Theorem 2.9 fails)

Fig. 6.2 Existence of power series solutions in $\mathbb{R}[[x]]$ for some polynomial equations

6.3 Puiseux Series and the Newton Polygon

If $m(x,y) \in \mathbb{K}[x,y]$ is an irreducible polynomial of degree d, then there usually will be also d roots of the univariate polynomial $m(0,y)$, and usually, each of these roots ζ can be extended to a power series solution $a(x)$ of $m(x,y) = 0$ with $a(0) = \zeta$. But we have seen in Fig. 6.2 that there are degenerate situations which deviate from this expected behavior.

Let us have a closer look at these degenerate situations. When we observed in Chap. 2 that the ring of power series is not closed under division, we came up with a generalized notion of series, Laurent series, with which division is always possible. Now we are faced with the situation that power series are not sufficiently fine-grained to express all the solutions of a bivariate algebraic equation. Even Laurent series are not enough for this purpose. So our goal now is to come up with some generalized notion of series, called *Puiseux series,* that comprises all the solutions of any algebraic equation.

While Laurent series are obtained from power series by allowing terms x^n with negative integers n to appear in the series, we will now extend further and allow even rational numbers as exponents. The precise construction is best explained with an example. Consider the polynomial

$$m(x,y) = (2+x+43x^2) - (2-10x+66x^2-162x^3)y$$
$$- (9x-53x^2+213x^3-405x^4)y^2 - (16x^2-120x^3+396x^4-540x^5)y^3$$
$$- (14x^3-120x^4+369x^5-405x^6)y^4 - (6x^4-54x^5+162x^6-162x^7)y^5$$
$$- (x^5-9x^6+27x^7-27x^8)y^6.$$

We want to describe all six solutions of the equation $m(x,y) = 0$ in terms of some series. We are willing to admit terms x^α with rational α in our series, but like for Laurent series, we insist that any series must have a starting point, i.e., our series should have the form $a(x) = a_\alpha x^\alpha + \cdots$ where the "$\cdots$" stands for a sum of terms $a_\beta x^\beta$ with $\beta > \alpha$. Potential values for α and a_α can be found by plugging $a_\alpha x^\alpha + \cdots$ into the equation:

$$(2+\cdots) - (2+\cdots)(a_\alpha x^\alpha + \cdots) - (9x+\cdots)(a_\alpha^2 x^{2\alpha} + \cdots)$$
$$- (16x^2+\cdots)(a_\alpha^3 x^{3\alpha} + \cdots) - (14x^3+\cdots)(a_\alpha^4 x^{4\alpha} + \cdots)$$
$$- (6x^4+\cdots)(a_\alpha^5 x^{5\alpha} + \cdots) - (x^5+\cdots)(a_\alpha^6 x^{6\alpha} + \cdots) \overset{!}{=} 0.$$

Admissible values for α must enable some cancellation on the left hand side. For example, the choice $\alpha = 3$ is not going to work out, because for this choice, all summands would have positive order, except for the first, which would have a lonely unmatched constant 2. There is no point in setting $\alpha = -3$ either, for then the last summand would produce a lonely term $x^5 x^{6\alpha} = x^{-13}$ with its exponent -13 being too low to be matched by any other term.

Cancellation can only happen if α is chosen in such a way that at there are at least two terms of minimal order on the left hand side. Since the minimal order term contributed by a summand $(x^j + \cdots)y^i$ of $m(x,y)$ is $x^{j+\alpha i}$, the numbers α we are interested in are those for which $j + \alpha i = n + \alpha k$ for at least two summands $(x^j + \cdots)y^i$ and $(x^n + \cdots)y^k$ of $m(x,y)$ and for which there is no other summand $(x^v + \cdots)y^u$ in $m(x,y)$ with $v + \alpha u < j + \alpha i$. These requirements can be restated as a simple geometric problem. For each term $x^j y^i$ appearing in $m(x,y)$, draw a vertical line from (i,j) upwards and determine the convex hull C of all these lines. We call C the *Newton polygon* of $m(x,y)$. It has the feature that the possible choices $\alpha = -(n-j)/(k-i)$ are precisely the negative slopes of the non-vertical edges (i,j)–(k,n) of C.

The Newton polygon for our present example is depicted in Fig. 6.3 on the left. We see that the only possible exponents are $\alpha = 0$ (by the edge connecting $(0,0)$ and $(1,0)$) and $\alpha = -1$ (by the edge connecting $(1,0)$ and $(6,5)$). For $\alpha = 0$, the equation $m(x, a_\alpha x^\alpha + \cdots) = 0$ simplifies to

$$(2-2a_0)x^0 + \text{higher order terms} \overset{!}{=} 0,$$

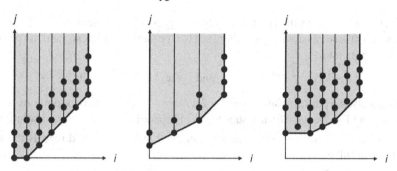

Fig. 6.3 Newton polygons used for solving $m(x,y) = 0$

and for $\alpha = -1$ we get

$$(-2a_{-1} - 9a_{-1}^2 - 16a_{-1}^3 - 14a_{-1}^4 - 6a_{-1}^5 - a_{-1}^6)x^{-1} + \text{higher order terms} \overset{!}{=} 0.$$

We can now determine a_0 and a_{-1} so as to let the least order terms cancel. This gives us $a_0 = 1$, and in the second case $a_{-1} = -1$ or $a_{-1} = -2$.

At this point we have evidence for the existence of three solutions starting like

$$1 + \cdots, \qquad -x^{-1} + \cdots, \qquad -2x^{-1} + \cdots,$$

respectively. No fractional exponents yet, and still only half as many (partial) solutions as we expect. But let us see what the next terms in these series are. We can determine them in very much the same way as we found the initial terms. In order to extend, say, the second partial solution by one term, we plug $y = -x^{-1} + a_\beta x^\beta$ into the equation $m(x,y) = 0$ and obtain

$$(16x + \cdots) + a_\beta^2 (8x^{2+2\beta} + \cdots) + a_\beta^4 (x^{3+4\beta} + \cdots) - a_\beta^6 (x^{5+6\beta} + \cdots) \overset{!}{=} 0.$$

The possible choices for β can be read off from the Newton polygon of $m(x, -x^{-1} + y)$, which is shown in Fig. 6.3 in the middle. We find $\beta = -1/2$ and $\beta = -1$, the latter being discarded because we are working on the solution with $\alpha = -1$ and we seek $\beta > \alpha$. Setting $\beta = -1/2$ in the previous equation leads to

$$(16 + 8a_\beta^2 + a_\beta^4)x + \text{higher order terms} \overset{!}{=} 0.$$

The coefficient of x vanishes if and only if $a_\beta = 2i$ or $a_\beta = -2i$. Two effects can be observed here: First, the partial solution $-x^{-1} + \cdots$ splits in this step into two different partial solutions

$$-x^{-1} + 2ix^{-1/2} + \cdots \quad \text{and} \quad -x^{-1} - 2ix^{-1/2} + \cdots,$$

and secondly, we now encounter terms with fractional exponents. The other two partial solutions do not split, but both have a unique successor term. They extend to

$$1 + x + \cdots \quad \text{and} \quad -2x^{-1} - 1 + \cdots,$$

respectively. Still we do not see six different solutions, so let us determine the next term of, say, $-x^{-1} + 2ix^{-1/2} + \cdots$. The Newton polygon of $m(x, -x^{-1} + 2ix^{-1/2} + y)$

is given in Fig. 6.3 on the right. The edges suggest the exponents $0, -1/2, -1$, of which only 0 is admissible. Again, there are two possible coefficients, and the partial solution splits into

$$-x^{-1} + 2ix^{-1/2} + 2ix^0 + \cdots \quad \text{and} \quad -x^{-1} + 2ix^{-1/2} - 2ix^0 + \cdots.$$

Analogously, the partial solution $-x^{-1} - 2ix^{-1/2} + \cdots$ splits into two solutions at the next step, and so we have finally succeeded in separating all six expected solutions.

The extraction of further terms can be continued indefinitely, and reveals the six solutions in full glory:

$$1 + x + 3x^2 + 7x^3 + 17x^4 + 47x^5 + 125x^6 + 333x^7 + \cdots,$$
$$-2x^{-1} - 1 - x - 3x^2 - 7x^3 - 17x^4 - 47x^5 - 125x^6 + \cdots,$$
$$-x^{-1} + 2ix^{-1/2} + 2i + 5ix^{1/2} + 7ix + \tfrac{59}{4}ix^{3/2} + \tfrac{87}{4}ix^2 + \tfrac{345}{8}ix^{5/2} + \cdots,$$
$$-x^{-1} - 2ix^{-1/2} + 2i - 5ix^{1/2} + 7ix - \tfrac{59}{4}ix^{3/2} + \tfrac{87}{4}ix^2 - \tfrac{345}{8}ix^{5/2} + \cdots,$$
$$-x^{-1} + 2ix^{-1/2} - 2i + 5ix^{1/2} - 7ix + \tfrac{59}{4}ix^{3/2} - \tfrac{87}{4}ix^2 + \tfrac{345}{8}ix^{5/2} + \cdots,$$
$$-x^{-1} - 2ix^{-1/2} - 2i - 5ix^{1/2} - 7ix - \tfrac{59}{4}ix^{3/2} - \tfrac{87}{4}ix^2 - \tfrac{345}{8}ix^{5/2} + \cdots.$$

In summary, we have solved a polynomial equation $m(x, y) = 0$ in terms of series of the form

$$a_\alpha x^\alpha + a_\beta x^\beta + a_\gamma x^\gamma + \cdots$$

term by term. We have plugged a symbolic first term $a_\alpha x^\alpha$ into the equation and obtained by inspection of the Newton polygon a finite number of exponents α for which the lowest order coefficient of $m(x, a_\alpha x^\alpha)$ is a nontrivial polynomial in a_α. By setting this polynomial to zero, we found the possible choices for the coefficients a_α. Higher order terms are obtained by substituting the known lower order terms into the equation and proceeding as described for the initial term, but taking only exponents into account which are greater than those already appearing among the known terms.

It is clear by construction that any series obtained by this procedure is actually a (formal) solution of the equation under consideration. Less obvious is whether the procedure will for every equation find a full set of solutions. This is asserted by Puiseux's theorem.

Theorem 6.2 (Puiseux) *If $\mathbb{K}$ is an algebraically closed field and $m(x, y) \in \mathbb{K}[x, y]$ is irreducible with $\deg_y m(x, y) = d$, then there exists a positive integer r and d distinct Laurent series $a_1(x), \ldots, a_d(x) \in \mathbb{K}((x))$ with $m(x^r, a_i(x)) = 0$ for $i = 1, \ldots, d$.*

After a formal substitution $x \mapsto x^{1/r}$, the Laurent series $a_i(x)$ in the theorem become objects of the form $a_i(x^{1/r})$ which involve fractional exponents and satisfy $m(x, a_i(x^{1/r})) = 0$. These are the series whose computation we discussed above. They are called *Puiseux series*. Observe that Puiseux series may involve fractional exponents, but not in an arbitrary fashion. The fractions appearing as exponents of a fixed Puiseux series must have a finite common denominator, so, for example $x^{-1} + x^{-1/2} + x^{-1/3} + x^{-1/4} + x^{-1/5} + \cdots$ is *not* a Puiseux series.

The proof of Theorem 6.2 proceeds by tracing the computation of the series solutions via the Newton polygon and showing that in every iteration there will be a suitable choice for the exponent. The technical details can be found in [59].

6.4 Closure Properties

Like other classes of power series that we discussed earlier, the class of algebraic power series provides properties which ensure that by applying certain operations to some algebraic power series we will not be kicked out of the class. As in the cases we considered before, this is useful for building up complicated algebraic power series from simple building blocks, as well as for proving identities.

For the class of algebraic power series, closure properties are most conveniently formulated and executed by means of resultants, so let us briefly review what they are and what they are good for. (Details can be found in [58].) Two univariate polynomials $p(x), q(x) \in \mathbb{K}[x]$ have a common factor in $\mathbb{K}[x]$ (viz. a common root in the algebraic closure $\bar{\mathbb{K}}$) if and only if $\gcd_x(p(x), q(x)) \neq 1$ and this is the case if and only if there exist non-zero $u(x), v(x) \in \mathbb{K}[x]$ with

$$u(x)p(x) + v(x)q(x) = 0 \tag{R}$$

and $\deg u(x) < \deg q(x)$ and $\deg v(x) < \deg p(x)$. The idea is to make an ansatz for the coefficients of $u(x)$ and $v(x)$ and turn (R) into a linear system of equations whose solution space is $\{0\}$ if and only if $p(x)$ and $q(x)$ are coprime (viz. they have no roots in common). Let $n = \deg p(x)$ and $m = \deg q(x)$ and write

$$p(x) = p_0 + p_1 x + \cdots + p_n x^n, \qquad q(x) = q_0 + q_1 x + \cdots + q_m x^m$$
$$u(x) = u_0 + u_1 x + \cdots + u_{m-1} x^{m-1}, \qquad v(x) = v_0 + v_1 x + \cdots + v_{n-1} x^{n-1}.$$

Then the linear system reads

$$
\begin{pmatrix}
p_0 & 0 & \cdots & 0 & q_0 & 0 & \cdots & \cdots & 0 \\
p_1 & p_0 & \ddots & \vdots & \vdots & \ddots & \ddots & & \vdots \\
\vdots & \ddots & \ddots & 0 & \vdots & & \ddots & \ddots & \vdots \\
\vdots & & \ddots & p_0 & q_{m-1} & & & \ddots & 0 \\
\vdots & & & p_1 & q_m & \ddots & & & q_0 \\
p_n & & & \vdots & 0 & \ddots & \ddots & & \vdots \\
0 & \ddots & & \vdots & \vdots & & \ddots & \ddots & q_{m-1} & \vdots \\
\vdots & \ddots & \ddots & \vdots & \vdots & & & \ddots & q_m & q_{m-1} \\
0 & \cdots & 0 & p_n & 0 & & \cdots & \cdots & 0 & q_m
\end{pmatrix}
\begin{pmatrix}
u_0 \\ u_1 \\ \vdots \\ \vdots \\ u_{m-1} \\ v_0 \\ v_1 \\ \vdots \\ v_{n-1}
\end{pmatrix}
= 0
$$

$$\underbrace{\qquad\qquad}_{m \text{ columns}} \quad \underbrace{\qquad\qquad}_{m \text{ columns}}$$

The matrix is called the *Sylvester matrix* of $p(x)$ and $q(x)$, and the *resultant* of $p(x)$ and $q(x)$, written $\mathrm{res}_x(p(x),q(x))$, is defined as the determinant of the Sylvester matrix. Since the nullspace of a square matrix is nontrivial if and only if its determinant is zero, we can record

$$\gcd(p(x),q(x)) = 1 \iff \mathrm{res}_x(p(x),q(x)) \neq 0.$$

The exciting feature of this equivalence is not that we can use it for finding out *whether* two polynomials have a common root (this we could also do with Euclid's algorithm), but that we can use it for finding out *when* two polynomials have a common root: If we have bivariate polynomials $p(x,y), q(x,y) \in \mathbb{K}[x,y]$, then $r(y) := \mathrm{res}_x(p(x,y),q(x,y))$ will be a polynomial in y alone, and the roots of $r(y) \in \mathbb{K}[y]$ in $\mathbb{K}$ are precisely the values we may substitute for y such that the leading coefficient of both $p(x,y)$ and $q(x,y)$ with respect to x vanishes or $p(x,y)$ and $q(x,y)$ have a common factor as polynomials in $\mathbb{K}[x]$. (Readers who have not seen this before are strongly advised to read the present paragraph once more.)

Theorem 6.3 *Let $a(x)$ and $b(x)$ be algebraic power series in $\mathbb{K}[[x]]$, and let $p(x,y)$ and $q(x,y)$ in $\mathbb{K}[x,y]$ be annihilating polynomials for $a(x)$ and $b(x)$, respectively. Then:*

1. *$a'(x)$ is an algebraic power series, and $\mathrm{res}_z(p(x,z), D_x p(x,z) + y D_z q(x,z))$ is an annihilating polynomial.*

2. *If $a(0) \neq 0$, then $1/a(x)$ is an algebraic power series, and $\mathrm{res}_z(p(x,z), yz - 1)$ is an annihilating polynomial.*

3. *$a(x) + b(x)$ is an algebraic power series, and $\mathrm{res}_z(p(x,y-z), q(x,z))$ is an annihilating polynomial.*

4. *$a(x)b(x)$ is an algebraic power series, and $\mathrm{res}_z(z^{\deg_y P} p(x,y/z), q(x,z))$ is an annihilating polynomial.*

5. *If $b(0) = 0$, then $a(b(x))$ is an algebraic power series, and $\mathrm{res}_z(p(z,y), q(x,z))$ is an annihilating polynomial.*

Proof. We prove item 3 and leave the others to the interested reader.

Let $r(x,y) = \mathrm{res}_z(p(x,y-z), q(x,z))$. First of all, from the definition of the resultant as the determinant of a matrix whose entries, in the present case, are elements of $\mathbb{K}(x)[y]$, it is clear that $r(x,y)$ belongs to $\mathbb{K}(x)[y]$.

If $q(x,z)$ and $p(x,z)$ in $\mathbb{K}(x,y)[z]$ are free of y, then all their factors are also free of y. A polynomial $u(x,z)$ in $\mathbb{K}(x,y)[z]$ divides $p(x,z)$ if and only if $u(x,y-z)$ divides $p(x,y-z)$. Every nontrivial factor $u(x,y-z)$ of $p(x,y-z)$ must therefore involve y nontrivially and can not be at the same time a factor of $q(x,z)$. Therefore $q(x,z)$ and $p(x,z-y)$ have no factors in common and therefore $r(x,y)$ cannot be zero identically.

To see, finally, that $r(x,a(x)+b(x)) = 0$, it suffices to observe that the polynomials

$$p(x,a(x)+b(x)-z) \quad \text{and} \quad q(x,z)$$

(viewed now as elements of $\mathbb{K}((x))[z]$) have a common root $b(x)$ in $\mathbb{K}((x))$, and so $a(x) + b(x)$ must be a root of $r(x,y)$. □

The annihilating polynomials provided by Theorem 6.3 are in general not minimal polynomials. For example, let $a(x), b(x) \in \mathbb{K}[[x]]$ be defined via

$$a(x)^2 - (x+4) = 0, \qquad a(0) = 2,$$
$$b(x)^2 - (x+1)^2(x+4) = 0, \qquad b(0) = 2,$$

and let $c(x) = a(x) + b(x)$. We have

$$\text{res}_z((y-z)^2 - (x+4), z^2 - (x+1)^2(x+4))$$
$$= y^4 - 2(x+4)(x^2 + 2x + 2)y^2 + x^2(x+2)^2(x+4)^2$$
$$= (y^2 - x^3 - 4x^2)(y^2 - x^3 - 8x^2 - 20x - 16).$$

This cannot be the minimal polynomial of $c(x)$, because it is not irreducible. But the minimal polynomial must be among its factors. It cannot be the first, because

$$c(x)^2 - x^3 - 4x^2 = 16 + 20x + 4x^2 + \cdots \neq 0.$$

So it must be the second.

Care must be applied with operations not covered by Theorem 6.3. In particular, if $a(x)$ is an algebraic power series, $\int_x a(x)$ need not be, as the example $a(x) = 1/(1-x)$ shows. Also the Hadamard product of two algebraic power series is in general no longer algebraic. An example is

$$a(x) = b(x) = \sqrt{1+x} = \sum_{n=0}^{\infty} \binom{1/2}{n} x^n,$$

for which we have

$$a(x) \odot b(x) = \sum_{n=0}^{\infty} \binom{1/2}{n}^2 x^n.$$

We will see at the end of the next section that this series cannot be algebraic.

6.5 The Tetrahedron for Algebraic Functions

This time the starting vertex for going through the tetrahedron obviously is the generating function corner, because the defining property of the objects we are studying right now is explicitly formulated on the power series level and not on the level of sequences. The question is: what does algebraicity of a power series imply for its coefficient sequence?

As for summation, we have the closure property that if $(a_n)_{n=0}^{\infty}$ is the coefficient sequence of an algebraic power series $a(x)$, then the generating function $a(x)/(1-x)$ of $(\sum_{k=0}^{n} a_k)_{n=0}^{\infty}$ is again algebraic. Indeed, if $m(x,y)$ is the minimal polynomial of $a(x)$, then then $0 = m(x, a(x)) = m(x, (1-x)/(1-x)a(x))$, so $m(x, (1-x)y)$ is an annihilating polynomial for $a(x)/(1-x)$.

As for recurrence equations, it is not wise to start from the polynomial equation for the generating function. Expanding the powers $a(x)^k$ will lead to $(k-1)$-fold nested sums that are hardly useful for anything. In the next chapter we will see how the differential equation asserted by Theorem 6.1 gives rise to a linear recurrence equation with polynomial coefficients for the coefficient sequence of any algebraic power series. These are the recurrence equations of choice.

As for asymptotics, there is more to be said. We are in the fortunate situation that all algebraic power series are convergent in a neighborhood of the origin, so that we can actually regard them as analytic functions ("algebraic functions") and obtain information about the asymptotic growth of the coefficient sequences by means of singularity analysis. In order to do so, we need to (1) find out where the singularities are, and (2) find out how the algebraic function behaves in their neighborhood. In what follows, we assume that $\mathbb{K} = \mathbb{C}$, even where we use $\mathbb{K} = \mathbb{R}$ to gain geometric intuition.

If $m(x,y)$ is an irreducible polynomial of degree d in y, then for a particular choice $x = \xi$ there will typically be d different values $y = \zeta$ with $m(\xi, \zeta) = 0$. These d different values belong to the d different branches of the algebraic curve defined by $m(x,y)$, each of which gives rise to a particular analytic function in a neighborhood of ξ. For example, the algebraic curve shown in the upper left plot of Fig. 6.2 crosses the vertical axis in the two points $(0,0)$ and $(0, -16/27)$. In a neighborhood of either of them, the curve can be regarded as the graph of a function $x \mapsto y(x)$. In this example, the neighborhood can be extended indefinitely to the right, but not to the left. The blocker on the left is the position x where the curve has a vertical tangent. Such a point is called a branch point.

Branch points are one kind of singularities an algebraic function may have. They appear at points $x = \xi$ where there exists some $y = \zeta$ such that both $m(\xi, \zeta) = 0$ (meaning (ξ, ζ) should be on the curve) and $D_y m(x,y)|_{x=\xi, y=\zeta} = 0$ (meaning the curve should have a vertical tangent at this point). How can we find these points ξ? With a resultant! The branch points are precisely the roots of the polynomial $\mathrm{res}_y(m(x,y), D_y m(x,y)) \in \mathbb{K}[x]$.

The other kind of singularity arises for values $x = \xi$ where the degree of the minimal polynomial drops. This is obviously the case precisely for those $x = \xi$ where the leading coefficient of $m(x,y)$ with respect to y (which is a univariate polynomial in x) vanishes. Geometrically, this sort of singularity corresponds to branches with a vertical asymptote; the algebraic function has a pole there.

These are all the points where a singularity can occur: branch points and points with poles. Since both kinds are the roots of certain univariate polynomials, we can be certain that there are only finitely many of them, and we can even compute them explicitly. All the singularities of an algebraic function must be among these "critical points". But a particular algebraic function need not have a singularity for every critical point. A particular algebraic function corresponds to just one branch of an algebraic curve, and ξ being a critical point only means that some of the branches have a singularity at ξ. For the particular power series at hand, we have to determine which of the possible singularities do actually belong to the branch of the algebraic function determined by the power series. While this is not an easy thing to do in general, it can often be done by inspection in concrete examples.

Let us suppose now that $a(x) := \sum_{n=0}^{\infty} a_n x^n \in \mathbb{C}[[x]]$ is an algebraic power series and let $\rho > 0$ be its radius of convergence (in the sense of analysis). Let $U := \{z \in \mathbb{C} : |z| < \rho\}$ be the disc of radius ρ centered at the origin and let

$$A: U \to \mathbb{C}, \qquad A(z) := \sum_{n=0}^{\infty} a_n z^n.$$

This function will have singularities at the boundary of U. Let us suppose for simplicity that there is only one, and let us call it ζ. Then this singularity alone determines the asymptotics of $(a_n)_{n=0}^{\infty}$, and a precise asymptotic estimate can be obtained by canceling the singularity as sketched in Sect. 2.4.

If $m(x,y)$ is the minimal polynomial of $a(x)$, then we have $m(z,A(z)) = 0$ for all $z \in U$. In particular $m(z,A(z)) \to 0$ as $z \to \zeta$ in U. Substituting $z \mapsto \zeta(1-z)$, it follows that A behaves near ζ like a solution y of the equation $m(\zeta(1-x),y) = 0$ near zero. Among the Puiseux series solutions of this equation there must therefore be one series which describes the asymptotic behavior of A near ζ. Which of them it is, this is again not easy to determine in general, but can usually be found out by inspection on concrete examples. If we have

$$A(z) = \sum_{k=k_0}^{\infty} A_k (1 - z/\zeta)^{k/r}$$

for all $z \in U$ which are sufficiently close to ζ, then the asymptotics of $(a_n)_{n=0}^{\infty}$ is determined by the A_k and the coefficients in the expansion of the terms $(1 - z/\zeta)^{k/r}$ for $k/r \in \mathbb{Q} \setminus \mathbb{N}$. These coefficients are readily determined by the binomial theorem and the asymptotics of hypergeometric sequences:

$$[x^n](1-x/\zeta)^{k/r} = \binom{k/r}{n}(-\zeta)^{-n} \sim \frac{1}{\Gamma(-k/r)}\left(\frac{1}{\zeta}\right)^n n^{-1-k/r} \qquad (n \to \infty).$$

In terms of absolute value, these terms grow more quickly for smaller values of k. This suggests that the relevant index k for the overall asymptotics of $(a_n)_{n=0}^{\infty}$ is the smallest $k \geq k_0$ for which $k/r \notin \mathbb{N}$. Indeed, it can be shown that for this index k we have

$$a_n \sim \frac{A_k}{\Gamma(-k/r)}\left(\frac{1}{\zeta}\right)^n n^{-1-k/r} \qquad (n \to \infty).$$

Our derivation is heuristic to the extent that it does not contain rigorous analytic justification that summing up the infinitely many terms of the Puiseux series expansion of A does not mess up the estimate obtained for the individual terms. For a complete argument, we refer once more to the book of Flajolet and Sedgewick [21] and the references given there. The general treatment extends to the case of several singularities $\zeta_1,\ldots,\zeta_m$ on the boundary of the disc of convergence. In this case, the asymptotics is of the form

$$a_n \sim c_1 \left(\frac{1}{\zeta_1}\right)^n n^{-1-\alpha_1} + \cdots + c_m \left(\frac{1}{\zeta_m}\right)^n n^{-1-\alpha_m} \qquad (n \to \infty)$$

for certain constants $c_1,\ldots,c_m \in \mathbb{C}$ and $\alpha_1,\ldots,\alpha_m \in \mathbb{Q} \setminus \mathbb{N}$.

As an example, let

$$a(x) = \sum_{n=0}^{\infty} a_n x^n = x^3 - 2x^4 + 7x^5 - 20x^6 + 64x^7 - 200x^8 + 647x^9 + \cdots.$$

be the power series defined by $m(x,a(x)) = 0$ where

$$m(x,y) = x^3 + (x-1)(3x+1)y + 3xy^2 + y^3.$$

We want to determine the asymptotics of $(a_n)_{n=0}^{\infty}$. The curve defined by $m(x,y)$ is depicted in Fig. 6.4. Since $a(0) = 0$, we are interested in the branch passing through the origin.

To determine the possible locations of the singularities, we compute

$$\text{res}_y(m(x,y), D_y m(x,y)) = 108x^4 + 76x^3 - 21x^2 - 24x - 4$$
$$= (2x+1)^2(27x^2 - 8x - 4).$$

Hence we find that the curve defined by $m(x,y)$ has vertical tangents at the positions

$$-\tfrac{1}{2}, \quad \tfrac{2}{27}(2-\sqrt{31}), \quad \text{and} \quad \tfrac{2}{27}(2+\sqrt{31}).$$

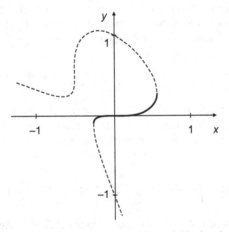

Fig. 6.4 Algebraic curve defining an analytic function in a neighborhood of $x = 0$

Closest to the origin on the branch we are working on is $\zeta := \frac{2}{27}(2 - \sqrt{31}) \approx -0.26$. (If we were working on the branch going through $(0,1)$, the limiting singularity would be $-1/2$.) On $U := \{z \in \mathbb{C} : |z| < |\zeta|\}$ we can therefore define the analytic function

$$A: U \to \mathbb{C}, \quad A(z) := \sum_{n=0}^{\infty} a_n z^n.$$

In order to clarify the behavior of A near ζ, consider the polynomial $m(\zeta(1-x), y)$. Its three Puiseux series solutions start as

$$-4\zeta + \tfrac{1}{81}(32 + 108\zeta)x - \tfrac{1}{59049}(4448 - 15012\zeta)x^2 + \cdots,$$
$$\tfrac{1}{2}\zeta + \tfrac{1}{3}\sqrt{2 + 2\zeta}\,x^{1/2} - \tfrac{1}{162}(32 - 135\zeta)x + \cdots,$$
$$\tfrac{1}{2}\zeta - \tfrac{1}{3}\sqrt{2 + 2\zeta}\,x^{1/2} - \tfrac{1}{162}(32 - 135\zeta)x + \cdots.$$

The first solution has no fractional exponents, is hence analytic at ζ and must therefore correspond to the branch going through $(0,1)$. The other two solutions correspond to the two branches attached to the branch point at ζ. Which of them corresponds to the branch going upwards depends on whether we choose to identify the formal term $x^{1/2}$ with the analytic function $+\sqrt{z}$ or $-\sqrt{z}$. In either case, we find that

$$A(z) = \tfrac{1}{2}\zeta + \tfrac{1}{3}\sqrt{2 + 2\zeta}\,\sqrt{1 - z/\zeta} + \cdots \qquad (z \to \zeta \text{ in } U),$$

and from the square root term in this expansion we obtain

$$a_n \sim \frac{\sqrt{2 + 2\zeta}}{3\Gamma(-1/2)}\zeta^{-n}n^{-1-1/2} = -\frac{\sqrt{2 + 2\zeta}}{6\sqrt{\pi}}\left(-\frac{2 + \sqrt{31}}{2}\right)^n n^{-3/2} \qquad (n \to \infty).$$

It is easy to do a cross check of this result by comparing the actual values of a_n and the asymptotic term for some large n. In fact, already for indices of moderate size, the asymptotics becomes apparent. For example, for $n = 25$ we have $a_{25} = 254927203828$ while the asymptotic estimate evaluates to $256256891882.55236\ldots$. The difference is already below 1%.

Singularity analysis not only provides us with a powerful means to determine the asymptotics of particular sequences, but it also gives rise to general structure theorems. For example, in the case of algebraic functions, since it is true *in general* that the asymptotics of a coefficient sequence is of the form $c\phi^n n^{-1-\alpha}$ for some $c, \phi \in \mathbb{C}$ and $\alpha \in \mathbb{Q} \setminus \mathbb{N}$, we can reject right away all power series whose coefficients have a different asymptotic behavior as transcendental. This applies in particular to the power series

$$\sum_{n=0}^{\infty} \binom{1/2}{n}^2 x^n,$$

because the formulas given in Chap. 5 imply

$$\binom{1/2}{n}^2 \sim \frac{1}{4\pi}n^{-3} \qquad (n \to \infty),$$

and there is no $\alpha \in \mathbb{Q} \setminus \mathbb{N}$ with $-3 = -1 - \alpha$.

6.6 Applications

Context Free Languages

A formal language $L \subseteq \Sigma^*$ is said to be context free if its words can be described by
a context free grammar. A context free grammar is a way of "generating" the words
of a language in a deduction scheme that employs letters taken from some auxiliary
alphabet N disjoint with Σ. The letters in N are called *nonterminal symbols*. The
starting point of a deduction is some fixed distinguished letter $S \in N$, the *starting
symbol*. The context free grammar consists of a set of rules of the form $A \to \alpha$ where
$A \in N$ and $\alpha \in (N \cup \Sigma)^*$, and the words of the language defined by the grammar
are those words $\omega \in \Sigma^*$ which can be obtained by starting from S and repeatedly
replacing letters $A \in N$ by words $\alpha \in (N \cup \Sigma)^*$ according to the rules $A \to \alpha$ of the
grammar, until all nonterminal symbols have disappeared.
As an example, take $N = \{S, B, N, D\}$ and $\Sigma = \{+, *, -, (,), x, 0, 1, \dots, 9\}$ and the
rules

(1)	$S \to (S B S)$	(5)	$B \to +$	(10)	$D \to 0$
(2)	$S \to (- S)$	(6)	$B \to -$	(11)	$D \to 1$
(3)	$S \to N$	(7)	$B \to *$		$\vdots$
(4)	$S \to x$	(8)	$N \to D N$	(19)	$D \to 9.$
		(9)	$N \to D$		

A typical derivation using this grammar is

$$S \overset{(1)}{\to} (S B S) \overset{(1)}{\to} (S B (S B S)) \overset{(7)}{\to} (S * (S B S)) \overset{(4)}{\to} (x * (S B S))$$

$$\overset{(1)}{\to} (x * ((S B S) B S)) \overset{(3)}{\to} (x * ((S B S) B N)) \overset{(9)}{\to} (x * ((S B S) B D))$$

$$\overset{(12)}{\to} (x * ((S B S) B 2)) \overset{(6)}{\to} (x * ((S B S) - 2)) \overset{(4)}{\to} (x * ((x B S) - 2))$$

$$\overset{(4)}{\to} (x * ((x B x) - 2)) \overset{(7)}{\to} (x * ((x * x) - 2)).$$

It is easy to see that all the words generated by this grammar are syntactically correct
encodings of univariate polynomials in x with integer coefficients.

The number of words of length n in a context free language can always be described by an algebraic power series. An annihilating polynomial can be obtained
from the deduction rules of the grammar. In the example above, if we denote
by $S(x), N(x), B(x), D(x)$ the power series whose coefficients carry the number of
words that can be generated according to the grammar by the nonterminal symbols
S, N, B, D, respectively, then we trivially have $B(x) = 3x$ and $D(x) = 10x$. For the
other two series, the deduction rules translate directly into the relations

$$N(x) = D(x) + D(x) N(x) \quad \text{and} \quad S(x) = x^2 S(x)^2 B(x) + x^3 S(x) + N(x) + x.$$

The first implies $N(x) = D(x)/(1 - D(x)) = 10x/(1 - 10x)$, and the second gives

$$(30x^4 - 3x^3)S(x)^2 + (10x^4 - x^3 - 10x + 1)S(x) + (10x^2 - 11x) = 0$$

Restricted Lattice Walks

The trick we applied for obtaining the generating function of Dyck paths in the beginning of this chapter works more generally for enumeration problems related to lattice walks that are restricted to some part of the lattice. It is known as the *kernel method*. We give another example for this method.

As in Sect. 4.6, consider lattice walks in $\mathbb{Z}^2$ starting in the origin $(0,0)$ and consisting of n steps where each single step can go either north-east ($\nearrow$) or south ($\downarrow$) or west ($\leftarrow$). We are now interested in the number of walks which never step below the horizontal axis and end at a point (i,j). If $a_{n,i,j}$ denotes this number, then we have again the recurrence

$$a_{n+1,i,j} = a_{n,i-1,j-1} + a_{n,i+1,j} + a_{n,i,j+1} \qquad (n,j \geq 0; i \in \mathbb{Z}),$$

but this time in addition to the initial condition

$$a_{0,i,j} = \begin{cases} 1 & \text{if } i = j = 0 \\ 0 & \text{otherwise} \end{cases}$$

we have the boundary condition $a_{n,i,j} = 0$ $(n \geq 0, i \in \mathbb{Z}, j < 0)$ reflecting the condition that we must not step below the horizontal axis. Taking this boundary condition into account, the recurrence now implies the functional equation

$$\frac{1}{t}\big(a(t,x,y) - 1\big) = xya(t,x,y) + \frac{1}{x}a(t,x,y) + \frac{1}{y}\big(a(t,x,y) - a(t,x,0)\big)$$

for the generating function $a(t,x,y) := \sum_{n=0}^{\infty} \big(\sum_{i,j} a_{n,i,j} x^i y^j\big) t^n$. After bringing this equation into the form

$$\big(xy - t(x^2y^2 + x + y)\big)a(t,x,y) = xy - txa(t,x,0),$$

we let the left hand side disappear by mapping

$$y \mapsto y(t,x) := \frac{x - t - \sqrt{(t-x)^2 - 4t^2x^3}}{2tx^2}$$

$$= t + \frac{1}{x}t^2 + \frac{x^3+1}{x^2}t^3 + \frac{3x^3+1}{x^3}t^4 + \cdots \in \mathbb{Q}(x)[[t]]$$

and obtain first

$$a(t,x,0) = y(t,x)/t = 1 + \frac{1}{x}t + \frac{x^3+1}{x^2}t^2 + \frac{3x^3+1}{x^3}t^3 + \frac{6x^3+2x^6+1}{x^4}t^4 + \cdots,$$

and secondly

$$a(t,x,y) = \frac{x(y - y(t,x))}{xy - t(x^2y^2 + x + y)}.$$

Setting $x = y = 1$ finally tells us that

$$a(t,1,1) = \frac{3t - 1 + \sqrt{1 - 2t - 3t^2}}{2t(1 - 3t)} = 1 + 2t + 5t^2 + 13t^3 + 35t^4 + \cdots$$

counts the total number of walks of the type we consider.

It can be shown more generally that the generating function for lattice walks starting at $(0,0)$, consisting of steps taken from a fixed step set $S \subseteq \{\leftarrow, \nwarrow, \uparrow, \nearrow, \rightarrow, \searrow, \downarrow, \swarrow\}$, and confined to some half-plane is always algebraic.

Legendre Polynomials

Consider the algebraic power series

$$\sum_{n=0}^{\infty} P_n(x)t^n := (1 - 2xt + t^2)^{-1/2} = 1 + xt + \tfrac{1}{2}(3x^2 - 1)t^2 + \tfrac{1}{2}(5x^3 - 3x)t^3$$

$$+ \tfrac{1}{8}(35x^4 - 30x^2 + 3)t^4 + \tfrac{1}{8}(63x^5 - 70x^3 + 15x)t^5 + \cdots \in \mathbb{Q}(x)[[t]].$$

We are interested in the coefficients $P_n(x)$. Using the binomial theorem, we can express them in terms of a sum. The calculation

$$(1 - 2xt + t^2)^{-1/2} = (1 + t(t - 2x))^{-1/2} = \sum_{j=0}^{\infty} \binom{-1/2}{j}(t - 2x)^j t^j$$

$$= \sum_{j=0}^{\infty}\sum_{k=0}^{j} \binom{-1/2}{j}\binom{j}{k}(-2x)^{j-k}t^{j+k}$$

$$\overset{n=j+k}{=} \sum_{n=0}^{\infty} \left(\sum_{k=0}^{\lfloor n/2 \rfloor} \binom{-1/2}{n-k}\binom{n-k}{k}(-2x)^{n-2k} \right) t^n$$

implies

$$P_n(x) = \sum_{k=0}^{\lfloor n/2 \rfloor} \binom{-1/2}{n-k}\binom{n-k}{k}(-2x)^{n-2k} \qquad (n \geq 0).$$

From this representation it is apparent that the $P_n(x)$ are polynomials in x of degree n. They are called the *Legendre polynomials*.

Zeilberger's algorithm applied to the sum representation of $P_n(x)$ delivers the recurrence

$$(n+2)P_{n+2}(x) - (3+2n)xP_{n+1}(x) + (n+1)P_n(x) = 0 \qquad (n \geq 0).$$

Using this recurrence, it can be shown inductively that

$$\int_{-1}^{1} P_n(x)P_m(x)\,dx = \begin{cases} 0 & \text{if } n \neq m \\ 2/(2n+1) & \text{if } n = m \end{cases},$$

meaning that the Legendre polynomials form a family of orthogonal polynomials for the constant weight function 1.

Like Chebyshev polynomials, Legendre polynomials enjoy a lot of useful properties that make them attractive for numerical analysts.

6.7 Problems

Problem 6.1 We have deduced from the equation $1 - C(x) + xC(x)^2 = 0$ the explicit representation $C(x) = \frac{1-\sqrt{1-4x}}{2x}$. What rules out the possibility $C(x) = \frac{1+\sqrt{1-4x}}{2x}$?

Problem 6.2 Following the reasoning in the proof of Theorem 6.1, derive a differential equation for $C(x) = \frac{1-\sqrt{1-4x}}{2x}$ starting from its minimal polynomial.

Problem 6.3 Recall from Problem 2.10 that the infinite continued fraction

$$K(x) := \cfrac{x}{1 - \cfrac{x}{1 - \cfrac{x}{1 - \cdots}}}$$

exists as a formal power series in $\mathbb{K}[[x]]$. Prove that $K(x) = xC(x)$, where $C(x)$ is the generating function of the Catalan numbers.

Problem 6.4 It follows from the proof of Theorem 6.1 that a minimal polynomial $m(x,y) \in \mathbb{K}(x)[y]$ of degree d gives rise to a homogeneous linear differential equation of order (at most) d. Prove analogously that a minimal polynomial of degree d also implies the existence of an *inhomogeneous* linear differential equation of order (at most) $d - 1$, i.e., show that whenever $a(x) \in \mathbb{K}[[x]]$ is such that $m(x, a(x)) = 0$ then there are polynomials $p_0(x), \ldots, p_{d-1}(x), q(x) \in \mathbb{K}[x]$, not all zero, such that

$$p_0(x)a(x) + p_1(x)a'(x) + \cdots + p_{d-1}a^{(d-1)}(x) = q(x).$$

Problem 6.5 Determine the first four terms of all three Puiseux series $a(x)$ satisfying

$$4a(x)^3 - 3(8x+1)a(x) + (8x^2 + 20x - 1) = 0.$$

Problem 6.6 Let $a(x) = \sum_{n=0}^{\infty} a_n x^n$ and $b(x) = \sum_{n=0}^{\infty} b_n x^n$ be two algebraic power series and let $(c_n)_{n=0}^{\infty}$ be the sequence starting like

$$a_0,\ b_0,\ a_1,\ b_1,\ a_2,\ b_2,\ a_3,\ b_3,\ \dots.$$

Show that $\sum_{n=0}^{\infty} c_n x^n$ is algebraic.

Problem 6.7 Determine the minimal polynomial of the algebraic power series

$$\sum_{n=0}^{\infty} (C_n + F_n) x^n,$$

where $(C_n)_{n=0}^{\infty}$ and $(F_n)_{n=0}^{\infty}$ are the sequences of Catalan and Fibonacci numbers, respectively.

Problem 6.8 Let

$$a(x) = \sum_{n=0}^{\infty} a_n x^n = \tfrac{1}{3}x + \tfrac{2}{27}x^2 + \tfrac{64}{2187}x^3 + \tfrac{32}{2187}x^4 + \cdots$$

be the power series satisfying

$$64a(x)^3 - 72a(x)^2 + (48x - 81)a(x) + (27x - 2x^2) = 0, \quad a(0) = 0.$$

Determine the asymptotics of $(a_n)_{n=0}^{\infty}$.

Problem 6.9 The hypergeometric series

$$_2F_1\left(\begin{array}{c} -1/3, -2/3 \\ 1/2 \end{array}\middle| x\right) = 1 + \tfrac{4}{9}x + \tfrac{8}{243}x^2 + \tfrac{64}{6561}x^3 + \tfrac{256}{59049}x^4 + \cdots$$

is algebraic. Find its minimal polynomial.

Is every hypergeometric series algebraic? Is every algebraic series hypergeometric?

Problem 6.10 The *diagonal* of a bivariate power series $a(x,y) = \sum_{n,k=0}^{\infty} a_{n,k} x^n y^k \in \mathbb{K}[[x,y]]$ is defined as $d(x) := \sum_{n=0}^{\infty} a_{n,n} x^n \in \mathbb{K}[[x]]$. It can be shown that whenever $a(x,y)$ is rational, then its diagonal $d(x)$ is algebraic. The goal of this problem is to confirm this theorem for the particular example

$$a(x,y) := \sum_{n=0}^{\infty} a_{n,k} x^n y^k := \frac{1}{1 - x - y + 2xy}.$$

1. Use the geometric series twice to determine a hypergeometric sum description for $a_{n,k}$ $(n,k \geq 0)$.
2. Set $n = k$ in this sum and apply Zeilberger's algorithm to find a recurrence for $(a_{n,n})_{n=0}^{\infty}$.
3. Use the recurrence to compute the first, say, twenty terms on the diagonal, and apply automated guessing to find a conjectural algebraic equation for $d(x) = \sum_{n=0}^{\infty} a_{n,n} x^n$.
4. Prove the conjectured equation.

Problem 6.11 Prove item 4 of Theorem 6.3.

Problem 6.12 Let $a(x)$ be an algebraic power series and suppose that the power series $b(x)$ is such that $a(b(x)) = x$. Show that $b(x)$ is algebraic.

Problem 6.13 Here are some ways to prove that $\exp(x) = \sum_{n=0}^{\infty} \frac{1}{n!} x^n$ is not algebraic.

1. Confirm that the coefficient sequence $(1/n!)_{n=0}^{\infty}$ is incompatible with the possible asymptotic growths of coefficient sequences of algebraic functions.
2. Suppose that $p(x,y) \in \mathbb{Q}[x,y]$ is the minimal polynomial of $\exp(x)$. Derive a contradiction by showing that differentiating $p(x,\exp(x))$ gives a smaller annihilating polynomial.
3. Suppose that $p(x,y) \in \mathbb{Q}[x,y]$ is the minimal polynomial of $\exp(x)$. Derive a contradiction by considering $\lim_{z \to \infty} p(z, e^z) z^u e^{vz}$ for suitably chosen $u, v \in \mathbb{Z}$.
4. Argue with Theorem 6.3 that if $\exp(x) \in \mathbb{C}[[x]]$ is algebraic then also $x/\sin(x)$ is algebraic. Then use that algebraic functions have at most finitely many singularities.
5. Show that for any $d \in \mathbb{N}$, if $p_0(x), \ldots, p_d(x) \in \mathbb{Q}[x]$ are such that

$$p_0(n) + p_1(n)\frac{1}{n!} + p_2(n)\frac{2^n}{n!} + \cdots + p_d(n)\frac{d^n}{n!} = 0 \quad (n \geq 0)$$

then $p_0(x) = p_1(x) = \cdots = p_d(x) = 0$. (*Hint:* Reuse the idea from part (*ii*) of the proof of Theorem 4.1.) Conclude that $\exp(x)$ cannot be algebraic.

Problem 6.14 Prove that $\sum_{n=0}^{\infty} H_n x^n$ is not algebraic.
(*Hint:* First show that $\log(1+x)$ is not algebraic by combining the results of the previous two problems.)

Problem 6.15 Let $a(x) = \sum_{n=0}^{\infty} a_n x^n$. Prove or disprove:

1. If $p(x, a(x)) = 0$ for some nonzero polynomial $p(x,y)$, then $q(n, a_n) = 0$ $(n \geq 0)$ for some nonzero polynomial $q(x,y)$.
2. If $q(n, a_n) = 0$ $(n \geq 0)$ for some nonzero polynomial $q(x,y)$, then $p(x, a(x)) = 0$ for some nonzero polynomial $p(x,y)$.

Problem 6.16 The grammar $\{S \to (S), S \to SS, S \to \varepsilon\}$ for the alphabet $\Sigma = \{(,)\}$ and the single nonterminal symbol S describes the language consisting of all correctly bracketed words:

$$\varepsilon, \; (), \; ()\,(), \; (()), \; ()\,()\,(), \; (())\,(), \; ()\,(()), \; ((())), \; (()\,()), \ldots$$

This language obviously does not contain any words of odd length. How many words of length $2n$ does it contain?

Problem 6.17 Determine the generating function for the number of lattice walks confined to the upper half plane, starting in the origin, and consisting of exactly n steps that go either up ($\uparrow$), down ($\downarrow$), left ($\leftarrow$) or right ($\rightarrow$).

Problem 6.18 Prove the following alternative sum representations of Legendre polynomials by computing a recurrence equation for the sum (with Zeilberger's algorithm) and checking an appropriate number of initial values:

$$P_n(x) = 2^{-n} \sum_{k=0}^{n} \binom{n}{k}^2 (x-1)^{n-k}(x+1)^k = \sum_{k=0}^{n} \binom{n}{k}\binom{n+k}{k}\left(\frac{x-1}{2}\right)^k.$$

Chapter 7
Holonomic Sequences and Power Series

We are close to the end, but there is one more class that we want to present. This class is much wider than the classes discussed so far. It contains all the classes considered before, as well as many additional objects. It is interesting mainly for two reasons. First, because it covers a great percentage of the most widely used special functions in physics as well as many combinatorial sequences arising in applications. Secondly, because also for this large class there are algorithms for answering the most urging questions about its elements.

7.1 Harmonic Numbers

Harmonic numbers emerge in the analysis of the Quicksort algorithm, as we have seen at the beginning of the book. They also arise in numerous other contexts in enumerative combinatorics, in probability theory, or in number theory. They also have some properties which are interesting in their own right, being related even to some of the most famous open problems in mathematics altogether.

In several respects, the sequence $(H_n)_{n=0}^{\infty}$ of harmonic numbers resembles the natural logarithm. These two quantities are connected not only through the asymptotic estimate $H_n \sim \log(n)$ $(n \to \infty)$ or through the generating function

$$\sum_{n=0}^{\infty} H_n x^n = \frac{1}{1-x} \log \frac{1}{1-x},$$

but they also show a somewhat analogous behavior with respect to forward difference and differentiation:

$$\Delta (H_n)_{n=0}^{\infty} = \left(\tfrac{1}{n+1} \right)_{n=0}^{\infty} \quad \longleftrightarrow \quad D_t \log(t) = \tfrac{1}{t}.$$

This latter analogy has the consequence that many sums involving harmonic numbers can be evaluated in a similar fashion as integrals involving logarithms. For

M. Kauers, P. Paule, *The Concrete Tetrahedron*
© Springer-Verlag/Wien 2011

example, using the rule for summation by parts (Problem 3.12 in Chap. 3), we find

$$\underbrace{\sum_{k=0}^{n} H_k}_{u_k \ \Delta v_k} = \underbrace{\sum_{k=0}^{n} H_k}_{} \ 1 \ = \underbrace{H_{n+1}}_{u_{n+1}} \underbrace{(n+1)}_{v_{n+1}} - \underbrace{H_0}_{u_0} \underbrace{0}_{v_0} - \sum_{k=0}^{n} \underbrace{\frac{1}{k+1}}_{\Delta u_k} \underbrace{(k+1)}_{v_{k+1}}$$

$$= (n+1)H_{n+1} - (n+1) \qquad (n \geq 0).$$

Observe how closely this calculation matches the standard integral evaluation

$$\int \underbrace{\log(t)}_{u} \, dt = \int \underbrace{\log(t)}_{u} \underbrace{1}_{v'} \, dt = \underbrace{\log(t)t}_{u \ v} - \int \underbrace{\frac{1}{t}}_{u'} \underbrace{t}_{v} \, dt = t\log(t) - t.$$

Such analogies apply often, albeit not always as literally as here. The summation analogue of

$$\int \frac{1}{t} \log(t) \, dt = \frac{1}{2} \log(t)^2$$

illustrates that summation can sometimes be more subtle than integration: The identity

$$\sum_{k=1}^{n} \frac{1}{k} H_k = \frac{1}{2} \left(H_n^2 + \sum_{k=1}^{n} \frac{1}{k^2} \right) \qquad (n \geq 0),$$

which can also be found by summation by parts, contains in addition to the expected term a new sum which cannot be simplified any further. This sum defines the *harmonic numbers of second order.* More generally, for integers $r \geq 1$ we define

$$H_n^{(r)} := \sum_{k=1}^{n} \frac{1}{k^r} \qquad (n \geq 0),$$

and call $(H_n^{(r)})_{n=0}^{\infty}$ the sequence of *harmonic numbers of r-th order.* Generalizing further, we may allow several integer parameters $r_1, \ldots, r_m \geq 1$ and consider the nested sums

$$H_n^{(r_1, \ldots, r_m)} := \sum_{k_1=1}^{n} \sum_{k_2=1}^{k_1} \cdots \sum_{k_m=1}^{k_{m-1}} \frac{1}{k_1^{r_1} k_2^{r_2} \cdots k_m^{r_m}}.$$

In this notation, our previous result reads

$$H_n^{(1,1)} = \tfrac{1}{2} \left(H_n^2 + H_n^{(2)} \right) \qquad (n \geq 0).$$

It is a special case of the general identity

$$H_n^{(u,v)} + H_n^{(v,u)} = H_n^{(u)} H_n^{(v)} + H_n^{(u+v)} \qquad (n \geq 0)$$

which holds for arbitrary $u, v \geq 1$, and which, in a certain sense, describes all the relations that hold among the sequences $(H_n^{(u,v)})_{n=0}^{\infty}$ and $(H_n^{(w)})_{n=0}^{\infty}$ for $u, v, w \geq 1$. In contrast to the sequence $(H_n)_{n=0}^{\infty}$, which diverges for $n \to \infty$, the higher order harmonic numbers $(H_n^{(r)})_{n=0}^{\infty}$ $(r > 1)$ all converge to a finite limit. Also the sequences $(H_n^{(r_1,\ldots,r_m)})_{n=0}^{\infty}$ all converge to a finite limit whenever $r_1 > 1$. The limiting values are commonly denoted by the symbol ζ:

$$\zeta(r_1,\ldots,r_m) := \lim_{n \to \infty} H_n^{(r_1,\ldots,r_m)} \qquad (r_1 > 1, \; r_2,\ldots,r_m \geq 1).$$

The univariate version $(m = 1)$ is known as the Riemann zeta function and plays a prominent role in analytic number theory.

In spite of several centuries of mathematical research, the nature of the numbers $\zeta(r_1,\ldots,r_m)$ is still not well understood. We do know that there are further relations in addition to such relations as

$$\zeta(u,v) + \zeta(v,u) = \zeta(u)\zeta(v) + \zeta(u+v) \qquad (u, v > 1),$$

which directly follow from relations among the underlying generalized harmonic number sequences. For example, the identity

$$\zeta(2,1) = \zeta(3)$$

has no direct counterpart on the level of sequences. We also know (from Euler) that there are the closed form representations

$$\zeta(2n) = (-1)^{n+1} \frac{(2\pi)^{2n}}{2(2n)!} B_{2n} \qquad (n \geq 1),$$

where B_n is the n-th Bernoulli number. But we do not know, for instance, whether $\zeta(3)$ can be expressed in terms of common numbers such as π or e. It can definitely not be expressed as a fraction $p/q \in \mathbb{Q}$, as was proven only in 1978. No proof is known for the irrationality of $\zeta(n)$ for any other odd integer n. It is commonly believed (and there is strong empirical evidence supporting such believe) that all the numbers $\zeta(n)$ $(n \in \mathbb{N} \setminus \{0\})$ are transcendental. The point is that nobody knows how to formally prove this.

7.2 Equations with Polynomial Coefficients

The reader who has attentively followed the exercises of the previous chapters will have observed that the harmonic numbers $(H_n)_{n=0}^{\infty}$ are not a polynomial sequence, not a C-finite sequence, not a hypergeometric sequence, and not the coefficients of an algebraic series. Yet they do satisfy the simple second order recurrence equation

$$(n+2)H_{n+2} - (2n+3)H_{n+1} + (n+1)H_n = 0 \qquad (n \geq 0).$$

This recurrence renders the harmonic numbers *holonomic*.

We define a sequence $(a_n)_{n=0}^{\infty}$ to be *holonomic* (of order r and degree d) if there exist polynomials $p_0(x), \ldots, p_r(x) \in \mathbb{K}[x]$ of degree at most d with $p_0(x) \neq 0 \neq p_r(x)$ such that

$$p_0(n)a_n + p_1(n)a_{n+1} + \cdots + p_r(n)a_{n+r} = 0$$

for all $n \in \mathbb{N}$ with $p_r(n) \neq 0$. Such an equation is called a *holonomic recurrence* of order r and degree d. The sequence $(a_n)_{n=0}^{\infty}$ is uniquely determined by such a recurrence once we are given r initial values $a_0, \ldots, a_{r-1}$ and the values a_n for the (finitely many) indices n where $p_r(n) = 0$. *Nota bene:* a holonomic recurrence of order r can have more than r linearly independent solutions if $p_r(x)$ has non-negative integer roots.

A power series $a(x) \in \mathbb{K}[[x]]$ is called *holonomic* (of order r and degree d) if there exist polynomials $q_0(x), \ldots, q_r(x) \in \mathbb{K}[x]$ of degree at most d with $q_0(x) \neq 0 \neq q_r(x)$ such that

$$q_0(x)a(x) + q_1(x)D_x a(x) + \cdots + q_r(x)D_x^r a(x) = 0.$$

Such an equation is called a *holonomic differential equation* of order r and degree d. Holonomic sequences and holonomic power series are related as follows.

Theorem 7.1 *Let $a(x) = \sum_{n=0}^{\infty} a_n x^n \in \mathbb{K}[[x]]$ be a formal power series.*

1. *If the formal power series $a(x)$ is holonomic of order r and degree d, then the coefficient sequence $(a_n)_{n=0}^{\infty}$ is holonomic of order at most $r + d$ and degree at most r.*

2. *If the coefficient sequence $(a_n)_{n=0}^{\infty}$ is holonomic of order r and degree d, then the formal power series $a(x)$ is holonomic of order at most d and degree at most $r + d$.*

Proof. We show part 1. The proof of part 2 is Problem 7.5.

Suppose that $a(x)$ satisfies the holonomic differential equation

$$q_0(x)a(x) + q_1(x)D_x a(x) + \cdots + q_r(x)D_x^r a(x) = 0$$

for some polynomials $q_j(x) = q_{j,0} + q_{j,1}x + \cdots + q_{j,d}x^d \in \mathbb{K}[x]$ of degree at most d. The definition of differentiation and multiplication of formal power series directly implies

$$x^i D_x^j \sum_{n=0}^{\infty} a_n x^n = \sum_{n=i}^{\infty} (n-i+1)^{\bar{j}} a_{n+j-i} x^n \qquad (i, j \in \mathbb{N}).$$

Therefore

$$0 = \sum_{i=0}^{d} \sum_{j=0}^{r} q_{j,i} x^i D_x^j a(x) = \sum_{i=0}^{d} \sum_{j=0}^{r} q_{j,i} \left(\sum_{n=i}^{\infty} (n-i+1)^{\bar{j}} a_{n+j-i} x^n \right)$$

$$= \sum_{n=d}^{\infty} \left(\sum_{i=0}^{d} \sum_{j=0}^{r} q_{j,i}(n-i+1)^{\bar{j}} a_{n+j-i} \right) x^n + p(x)$$

for some polynomial $p(x)$ of degree less than d accounting for the necessary adjustments of initial values. Coefficient comparison with respect to x gives

$$\sum_{i=0}^{d}\sum_{j=0}^{r}q_{j,i}(n-i+1)^{\bar{j}}a_{n+j-i}=0 \qquad (n\geq d).$$

Substituting $n\mapsto n+d$, then $i\mapsto d-i$ and then setting $k=i+j$ turns this equation into

$$\sum_{k=0}^{d+r}\underbrace{\left(\sum_{j=\max(0,k-d)}^{\min(r,k)}q_{j,d+j-k}(n+k-j+1)^{\bar{j}}\right)}_{=:p_k(n)}a_{n+k}=0 \qquad (n\geq 0).$$

With the polynomials $p_0(x),\ldots,p_{d+r}(x)\in\mathbb{K}[x]$ defined as indicated by the brace, we have

$$p_0(n)a_n+p_1(n)a_{n+1}+\cdots+p_{d+r}(n)a_{n+(d+r)}=0 \qquad (n\geq 0).$$

This is the desired holonomic recurrence for $(a_n)_{n=0}^{\infty}$. $\qquad\qquad\square$

The theorem is constructive in the sense that it not only asserts the existence of certain differential or recurrence equations, but its proof also tells us how to obtain them. For example, if $a_n=n!$ $(n\geq 0)$, then the sequence $(a_n)_{n=0}^{\infty}$ is holonomic because of the first order recurrence equation

$$a_{n+1}-(n+1)a_n=0 \qquad (n\geq 0).$$

Therefore its generating function $a(x)=\sum_{n=0}^{\infty}n!x^n$ must satisfy a differential equation. Indeed,

$$x^2a''(x)+(3x-1)a'(x)+a(x)=0.$$

The class of holonomic sequences (viz. formal power series) encompasses all the classes considered in earlier chapters: polynomial sequences, C-finite sequences, and hypergeometric sequences obviously satisfy a holonomic recurrence equation, and algebraic series satisfy holonomic differential equations by Theorem 6.1. In addition, there are sequences and power series which are holonomic but do not belong to any of the classes previously considered (for example: harmonic numbers). Further holonomic objects can be obtained from the following closure properties.

Theorem 7.2 *Let* $a(x)=\sum_{n=0}^{\infty}a_nx^n\in\mathbb{K}[[x]]$ *and* $b(x)=\sum_{n=0}^{\infty}b_nx^n\in\mathbb{K}[[x]]$ *be holonomic. Then:*

1. *Linear combinations* $\alpha a(x)+\beta b(x)$ $(\alpha,\beta\in\mathbb{K}$ *fixed) are holonomic.*
2. *Cauchy product* $a(x)b(x)$ *and Hadamard product* $(a_nb_n)_{n=0}^{\infty}$ *are holonomic.*
3. *Derivative* $a'(x)$ *and forward shift* $(a_{n+1})_{n=0}^{\infty}$ *are holonomic.*

4. *Integral $\int_x a(x)$ and indefinite sum $(\sum_{k=0}^{n} a_k)_{n=0}^{\infty}$ are holonomic.*
5. *If $b(x)$ is algebraic and $b(0) = 0$, then $a(b(x))$ is holonomic.*
6. *$(a_{\lfloor un+v \rfloor})_{n=0}^{\infty}$ is holonomic for every $u, v \in \mathbb{Q}$ with $u, v \geq 0$.*

Proof. The proof of parts 1–4 and 6 resemble the proofs of C-finite closure properties (Theorem 4.2); part 5 is proven by the same argument as Theorem 6.1. As an illustration, we demonstrate the closure under Cauchy product in some detail.
If $p_0(x), \ldots, p_r(x) \in \mathbb{K}[x]$ with $p_r(x) \neq 0$ are such that

$$p_0(x)a(x) + p_1(x)D_x a(x) + \cdots + p_r(x)D_x^r a(x) = 0$$

then

$$D_x^r a(x) = -\frac{p_0(x)}{p_r(x)}a(x) - \frac{p_1(x)}{p_r(x)}D_x a(x) - \cdots - \frac{p_{r-1}(x)}{p_r(x)}D_x^{r-1}a(x).$$

Repeatedly differentiating this equation while always replacing any newly emerging term $D_x^r a(x)$ by the right hand side, it follows that for every $i \geq 0$ there exist rational functions $q_{0,i}(x), \ldots, q_{r-1,i}(x) \in \mathbb{K}(x)$ with

$$D_x^i a(x) = q_{0,i}(x)a(x) + q_{1,i}(x)D_x a(x) + \cdots + q_{r-1,i}(x)D_x^{r-1}a(x).$$

In other words, all derivatives $D_x^i a(x)$ ($i \geq 0$ fixed) belong to the $\mathbb{K}(x)$-vector space $A \subseteq \mathbb{K}((x))$ of dimension at most r generated by the power series $a(x), \ldots, D_x^{r-1}a(x)$. Analogously, if $b(x)$ is holonomic of order s, then the derivatives $D_x^j b(x)$ ($j \geq 0$ fixed) all belong to the $\mathbb{K}(x)$-vector space $B \subseteq \mathbb{K}((x))$ of dimension at most s generated by the power series $b(x), \ldots, D_x^{s-1}b(x)$.
Now let $c(x) = a(x)b(x)$. The derivatives $D_x^k c(x)$ ($k \geq 0$ fixed) all belong to the $\mathbb{K}(x)$-vector space $C = A \otimes B \subseteq \mathbb{K}((x))$ generated by the mutual products

$$a(x)b(x), \qquad (D_x a(x))b(x), \qquad \ldots, \qquad (D_x^{r-1}a(x))b(x),$$
$$a(x)(D_x b(x)), \quad (D_x a(x))(D_x b(x)), \quad \ldots, \quad (D_x^{r-1}a(x))(D_x b(x)),$$

$$\vdots$$

$$a(x)(D_x^{s-1}b(x)), (D_x a(x))(D_x^{s-1}b(x)), \ldots, (D_x^{r-1}a(x))(D_x^{s-1}b(x)).$$

The dimension of C is at most rs, so any $rs+1$ power series of the form $D_x^k c(x)$ ($k \geq 0$ fixed) must be linearly dependent. In particular, the power series $c(x), \ldots, D_x^{rs}c(x) \in C$ must be linearly dependent over $\mathbb{K}(x)$. This means that there must be rational functions $u_0(x), \ldots, u_{rs}(x) \in \mathbb{K}(x)$, not all zero, such that

$$u_0(x)c(x) + u_1(x)D_x c(x) + \cdots + u_{rs}(x)D_x^{rs}c(x) = 0.$$

Clearing denominators yields a holonomic differential equation of order at most rs for $c(x)$. $\square$

By this theorem, it is often a matter of seconds to realize that a power series given via a complicated-looking expression is holonomic. For instance, the power series

$$a(x) := \sum_{n=0}^{\infty} a_n x^n := \sum_{n=0}^{\infty} \left(\sum_{k=1}^{n} k! \right) x^n + \int_x \frac{\exp(1 - \sqrt{1+x}) \log(1-x)}{1 - x - x^2}$$

$$= x + \tfrac{5}{2}x^2 + \tfrac{26}{3}x^3 + \tfrac{389}{12}x^4 + \tfrac{36557}{240}x^5 + \tfrac{1116179}{1280}x^6 + \cdots \in \mathbb{Q}[[x]]$$

is obviously holonomic. Indeed, it satisfies a holonomic differential equation of order 7 and degree 32, which, printed out explicitly, would consume about two pages in this book.

The differential equation satisfied by a holonomic power series is not unique. Of course, starting from a particular equation of order r and degree d, we can obtain higher order equations of the same degree by differentiating, and higher degree equations of the same order can be obtained by simply multiplying the equation by a power of x. This way, we can create equations with order i and degree j for every point (i, j) in the cone $(r, d) + \mathbb{N}^2$ rooted at (r, d). The set of all the pairs (i, j) for which there exists an equation of order i and degree j is a finite union of such cones. For the example series $a(x)$ above, this set is visualized in Fig. 7.1. Equations of minimal order tend to have high degree and equations of minimal degree tend to have high order.

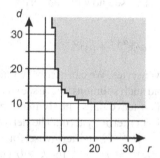 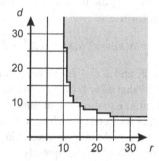

Fig. 7.1 Orders r and degrees d of possible holonomic differential equations (left) and recurrence equations (right) satisfied by $a(x)$ and $(a_n)_{n=0}^{\infty}$, respectively.

Differential equations satisfied by a holonomic power series can be recovered by automated guessing from its initial terms, provided that sufficiently many of them are known (Sect. 2.6). And once a reasonable guess is available, it is often possible to formally prove its correctness by an independent argument. In contrast, it can be quite hard to prove that a power series is not holonomic. Of course, the fact that we do not find an equation for it does not prove anything, as it may just mean that we did not search hard enough. One criterion which sometimes comes in handy is the (nontrivial) fact that for a holonomic power series $a(x) \in \mathbb{K}[[x]]$ with $a(0) \neq 0$ we have that $1/a(x)$ is holonomic if *and only if* $a'(x)/a(x)$ is algebraic. This criterion implies, for example, that the Bernoulli numbers are not holonomic, for if they were, then $x/(\exp(x) - 1)$ would be holonomic, and then, by applying the criterion to the holonomic power series $a(x) = (\exp(x) - 1)/x$, the series

$$\frac{a'(x)}{a(x)} = 1 - \frac{1}{x} + \frac{1}{\exp(x) - 1}$$

would be algebraic, which is not the case, as follows from Problem 6.13.

7.3 Generalized Series Solutions

The set of all power series solutions of a fixed holonomic differential equation forms a vector space over the ground field $\mathbb{K}$, for if two power series $a(x), b(x) \in \mathbb{K}[[x]]$ are solutions of a certain holonomic differential equation, then so is also $\alpha a(x) + \beta b(x)$ for any choice $\alpha, \beta \in \mathbb{K}$. Generically, for a differential equation of order r we expect the solution set to be a vector space of dimension r. But like in Chap. 6, we must be aware of degenerate situations in which there do not exist as many linearly independent power series solutions as the order of the equation suggests. Laurent series in general will not suffice either, and also Puiseux series are not always enough. Once more, we need a more generalized notion of series.

Before discussing the most general situation, let us see how to find series solutions $a(x)$ of the form

$$a(x) = x^\alpha \bar{a}(x) = a_0 x^\alpha + a_1 x^{\alpha+1} + a_2 x^{\alpha+2} + a_3 x^{\alpha+3} + \cdots$$

where $\alpha \in \mathbb{K}$ and $\bar{a}(x) \in \mathbb{K}[[x]]$ is a formal power series. We can assume without loss of generality that $a_0 = 1$. (Why?) In order to find such solutions, we first extract from the differential equation constraints on α which leave only finitely many possible values. Then we determine the coefficients of $\bar{a}(x)$ term by term. The derivative of x^α for $\alpha \in \mathbb{K}$ is defined as usual via $D_x x^\alpha := \alpha x^{\alpha-1}$, and termwise for a series of the form $x^\alpha \bar{a}(x)$. Consequently, we have $D_x^k x^\alpha = \alpha^{\underline{k}} x^{\alpha-k}$ for $k \geq 0$ and so if $\bar{a}(x)$ is a power series, then $D_x^k x^\alpha \bar{a}(x) = \alpha^{\underline{k}} x^{\alpha-k} + \cdots$. Substituting the template $a(x) = x^\alpha \bar{a}(x)$ into the left hand side of a given equation

$$p_0(x)a(x) + p_1(x)D_x a(x) + \cdots + p_r(x)D_x^r a(x) = 0$$

gives some series in which the coefficient of the lowest order term depends polynomially on α but not on any of the coefficients $a_1, a_2, a_3, \ldots$. This polynomial is called the *indicial polynomial* of the equation. Since all the coefficients must be zero in order for $a(x)$ to be a solution of the equation, the relevant values of α are precisely the roots of the indicial polynomial.

As an example, consider the equation

$$(6 + 6x - 3x^2)a(x) - (10x - 3x^2 - 3x^3)a'(x) + (4x^2 - 6x^3 + 2x^4)a''(x) = 0.$$

Aiming at solutions of the form $a(x) = x^\alpha \bar{a}(x)$ where $\bar{a}(x)$ is a power series with $\bar{a}(0) = 1$, we plug this solution template into the left hand side and obtain

$$(6 + \cdots)(x^\alpha + \cdots) - (10x + \cdots)(\alpha x^{\alpha - 1} + \cdots) + (4x^2 + \cdots)(\alpha(\alpha - 1)x^{\alpha - 2} + \cdots)$$
$$= (6 - 14\alpha + 4\alpha^2)x^\alpha + \cdots$$

There are only higher order terms hidden in the dots, so the only possible choices for α are the roots $1/2$ and 3 of the indicial polynomial $6 - 14\alpha + 4\alpha^2$. For either choice, we can next determine the series coefficients $a_1, a_2, a_3, \ldots$ by again making an ansatz, going into the equation, and comparing coefficients to zero. For example, the first terms of the series solution with $\alpha = 1/2$ are obtained by observing that

$$(6 + 6x - 3x^2)(x^{1/2} + a_1 x^{3/2} + a_2 x^{5/2} + a_3 x^{7/2} + \cdots)$$
$$- (10x - 3x^2 - 3x^3)(\tfrac{1}{2}x^{-1/2} + \tfrac{3}{2}a_1 x^{1/2} + \tfrac{5}{2}a_2 x^{3/2} + \tfrac{7}{2}a_3 x^{5/2} + \cdots)$$
$$+ (4x^2 - 6x^3 + 2x^4)(-\tfrac{1}{4}x^{-3/2} + \tfrac{3}{4}a_1 x^{-1/2} + \tfrac{15}{4}a_2 x^{1/2} + \tfrac{35}{4}a_3 x^{3/2} + \cdots)$$
$$= (9 - 6a_1)x^{3/2} + (-2 + 6a_1 - 4a_2)x^{5/2} + (3a_1 - 9a_2 + 6a_3)x^{7/2} + \cdots$$

turns into zero for $a_1 = \tfrac{3}{2}$, $a_2 = \tfrac{7}{4}$, $a_3 = \tfrac{15}{8}$, etc. In this example, we can continue to compute as many terms as we like. For every index n we will encounter a linear equation which determines the value of a_n uniquely in terms of the lower order terms determined earlier. But this is not always so. If the indicial polynomial has two roots α_1, α_2 at integer distance, say $\alpha_1 = \alpha_2 - k$ for some $k \in \mathbb{N}$, then the k-th coefficient of the solution corresponding to α_1 is not determined by the differential equation. Problem 7.12 has an example for this phenomenon.

What is true in general is that every root α of the indicial polynomial gives rise to a series solution of the form $x^\alpha \bar{a}(x)$. But unlike in the example, the number of roots of the indicial polynomial can be less than the order of the differential equation. In this case, there are solutions which cannot be written in the form $x^\alpha \bar{a}(x)$ but only in terms of more general types of series. There are two possible reasons for the indicial polynomial to lack some roots. First, its degree may be smaller than the order of the equation, and secondly, its roots may not be pairwise distinct.

It turns out that multiple roots indicate solutions which can be expressed using a (formal) logarithm with derivation rule $D_x \log(x) = 1/x$. To an m-fold root α of the indicial polynomial, there correspond m (formal) solutions

$$x^\alpha \bar{a}_0(x),$$
$$\log(x)x^\alpha \bar{a}_0(x) + x^\alpha \bar{a}_1(x),$$
$$\vdots$$
$$\log(x)^{m-1}x^\alpha \bar{a}_0(x) + \cdots + \binom{m-1}{k}\log(x)^k x^\alpha \bar{a}_{m-k}(x) + \cdots + x^\alpha \bar{a}_m(x)$$

where $\bar{a}_0(x), \ldots, \bar{a}_{m-1}(x)$ are formal power series with $\bar{a}_0(0) = 1$ and $\bar{a}_1(0) = \cdots = \bar{a}_{m-1}(0) = 0$. The coefficients of $\bar{a}_0(x)$ can be determined term by term just as described before, and a similar calculation gives the first n coefficients of the i-th series

$\bar{a}_i(x)$ once the first n coefficients of the series $\bar{a}_0(x), \ldots, \bar{a}_{i-1}(x)$ are known. For example, the holonomic differential equation

$$(1 - 24x + 96x^2)a(x) + (15x - 117x^2 + 306x^3)a'(x) + (9x^2 - 54x^3)a''(x) = 0$$

has the indicial polynomial $(1 + 3\alpha)^2$, indicating two solutions, one of the form $x^{-1/3}\bar{a}_0(x)$ and the other of the form $\log(x)x^{-1/3}\bar{a}_0(x) + x^{-1/3}\bar{a}_1(x)$. The first coefficients of the first solution are readily found to be

$$x^{-1/3}\left(1 + x + \tfrac{8}{3}x^2 + \tfrac{148}{27}x^3 + \tfrac{289}{27}x^4 + \cdots\right).$$

The second solution will therefore have the form

$$\log(x)x^{-1/3}\left(1 + x + \tfrac{8}{3}x^2 + \tfrac{148}{27}x^3 + \tfrac{289}{27}x^4 + \cdots\right)$$
$$+ x^{-1/3}\left(0 + a_1 x + a_2 x^2 + a_3 x^3 + a_4 x^4 + \cdots\right)$$

for some rational numbers $a_1, a_2, a_3, \ldots$ which can be determined by plugging the ansatz into the differential equation and comparing coefficients of like powers of x. This uniquely implies $a_1 = 1$, $a_2 = -\tfrac{59}{12}$, $a_3 = -\tfrac{6503}{324}$, $a_4 = -\tfrac{148393}{2592}$, etc.

The case of multiple roots being settled, it remains to investigate the case where the indicial polynomial does not have maximal possible degree. A simple example for this situation is the first order equation

$$a(x) + x^2 a'(x) = 0.$$

An attempt to solve this equation in terms of a series $x^\alpha(1 + a_1 x + a_2 x^2 + \cdots)$ leads to the condition

$$(x^\alpha + \cdots) + x^2(\alpha x^{\alpha-1} + \cdots) = 1x^\alpha + \cdots \overset{!}{=} 0$$

which can never be satisfied. The problem is that the increase in degree caused by the multiplication by x^2 exceeds the decrease in degree caused by the differentiation $a'(x)$. The consequence is that only $a(x)$ contributes to the indicial polynomial, but since its coefficients are free of α, the indicial polynomial is a nonzero constant here.

The idea is now to induce the appearance of a nontrivial indicial polynomial by applying suitable substitutions to the equation. It turns out that this can be achieved using (formal) exponentials. This time we will exploit the formal derivation rule

$$D_x \exp(cx^k) = ckx^{k-1}\exp(cx^k)$$

to produce multiplicative factors x^k according to our needs. Repeated derivation gives

$$D_x^m \exp(cx^k) = e_m(x^k)x^{-m}\exp(cx^k) \qquad (m \geq 0),$$

where $e_m(x)$ is a certain polynomial in $\mathbb{Q}(c)[x]$ of degree m. For $k \in \mathbb{Q}$, we may regard $e_m(x^k)$ as a Puiseux series with finitely many terms only, a "Puiseux polynomial" so to speak, whose lowest order term will have the exponent mk if k is negative.

Starting from a holonomic equation

$$p_0(x)a(x) + p_1(x)D_x a(x) + \cdots + p_r(x)D_x^r a(x) = 0,$$

the substitution $a(x) = \exp(cx^k)\bar{a}(x)$ for a fixed rational number $k < 0$ and a fixed constant $c \in \mathbb{K}$ leads to a new equation

$$\bar{p}_0(x)\bar{a}(x) + \bar{p}_1(x)D_x\bar{a}(x) + \cdots + \bar{p}_r(x)D_x^r\bar{a}(x) = 0$$

where $\bar{p}_0(x), \ldots, \bar{p}_r(x)$ are Puiseux series in x, specifically:

$$\bar{p}_i(x) = p_i(x) + \cdots + \binom{i+j}{j}e_j(x^k)x^{-j}p_{i+j}(x) + \cdots + \binom{r}{r-i}e_{r-i}(x^k)x^{-(r-i)}p_r(x)$$

for $i = 0, \ldots, r$. For the exponent of their lowest order term, we find

$$\operatorname{ord}\bar{p}_i(x) = \min\{\operatorname{ord}p_i(x), \, k-1+\operatorname{ord}p_{i+1}(x), \, \ldots, \, (r-i)(k-1)+\operatorname{ord}p_r(x)\}$$

for $i = 0, \ldots, r$. We have to find all values for k such that the equation for $\bar{a}(x)$ has a nontrivial indicial polynomial. The nontriviality of the indicial polynomial translates into an optimization problem for the orders of the $\bar{p}_i(x)$, which, similar as in Sect. 6.3, admits a simple geometric reformulation. For each term x^j occurring in $p_i(x)$, draw a vertical line from (i, j) upwards and determine the convex hull C of all these lines. We call C the *Newton polygon* of the differential equation. If C has an edge with slope $\alpha > 1$, then we substitute $a(x) = \exp(cx^k)\bar{a}(x)$ with $k = 1 - \alpha$ and undetermined c and $\bar{a}(x)$ into the left hand side of the original equation. The coefficient of the lowest order term will then be a nontrivial polynomial in c whose nonzero roots are precisely the eligible values for c.

The differential equation for $\bar{a}(x)$ obtained in this way may still not have a non-trivial indicial polynomial, but iterating the process with increasing values of k will after finitely many steps lead to a differential equation whose Newton polygon has a line segment with slope $\alpha = 1$. Such a segment indicates that we have reached an equation which has a nontrivial indicial polynomial. A solution of the equation can then be found as described earlier. The accumulated substitutions correspond to a multiplicative factor of the form $\exp(u(x^{-1/r}))$ for some $r \in \mathbb{N}$ and $u(x) \in \mathbb{K}[x]$.

To see a concrete example, consider the equation

$$216(1+x+x^3)a(x) + x^3(36 - 48x^2 + 41x^4)a'(x) - x^7(6+6x-x^2+4x^3)a''(x) = 0.$$

Its Newton polygon is depicted in Fig. 7.2 on the left. There are two choices for k. The edge $(0,0)$–$(1,3)$ offers $k = -2$ while the edge $(1,3)$–$(2,7)$ offers $k = -3$. We select the latter, substitute $a(x) = \exp(cx^{-3})\bar{a}(x)$ into the left hand side of the equation, and get

$$(c(2+c) + \cdots)\bar{a}(x) + (\cdots)\bar{a}'(x) + (\cdots)\bar{a}''(x) = 0$$

where the dots hide higher order terms. The coefficient of the lowest order term is $c(c+2)$. For $c = -2$ the equation has still no nontrivial indicial polynomial, so we

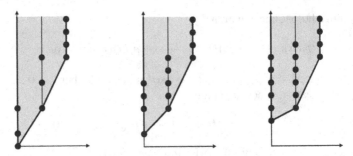

Fig. 7.2 Newton polygons for the example differential equation

iterate and consider the Newton polygon of the equation obtained for $c = -2$. It is shown in Fig. 7.2 in the middle. The edge $(1,3)$–$(2,7)$ was already used, so the only possible choice now is $(0,1)$–$(1,3)$, which corresponds to $k = -1$. Substituting $\bar{a}(x) = \exp(\bar{c}x^{-1})\bar{\bar{a}}(x)$ into the left hand side of the equation for $\bar{a}(x)$ gives an equation for $\bar{\bar{a}}(x)$ of the form

$$(36(1 - \bar{c})x + \cdots)\bar{\bar{a}}(x) + (\cdots)\bar{\bar{a}}'(x) + (\cdots)\bar{\bar{a}}''(x) = 0.$$

For $\bar{c} = 1$, the Newton polygon for this equation (Fig. 7.2 right) has an edge with slope 1, and indeed the equation has the nontrivial indicial polynomial $36(2 - \alpha)$ and a power series solution of order two with $1, -\frac{235}{6}, \frac{49375}{27}, \ldots$ as its first nonzero coefficients. It follows that the original differential equation has a solution of the form

$$a(x) = x^2 \exp(-2x^{-3} + x^{-1})\left(1 - \frac{235}{6}x + \frac{49375}{27}x^2 + \cdots\right).$$

Starting out with the edge $(0,0)$–$(1,3)$ leads analogously to the second solution

$$a(x) = x^{-2} \exp(3x^{-2})\left(1 + \frac{3}{4}x^2 - \frac{19}{18}x^3 + \frac{709}{288}x^4 + \cdots\right).$$

It is shown in [30] that the procedure described here works not only in this example but for any holonomic differential equation. In particular:

Theorem 7.3 *If $\mathbb{K}$ is algebraically closed, then every holonomic differential equation of order r has r linearly independent solutions of the form*

$$x^\alpha \exp(u(x^{-1/s}))\left(a_0(x) + \log(x)a_1(x) + \cdots + \log(x)^m a_m(x)\right)$$

where $s \in \mathbb{N} \setminus \{0\}$, $u(x) \in \mathbb{K}[x]$, $u(0) = 0$, $\alpha \in \mathbb{K}$, $m \in \mathbb{N}$, $a_0(x), \ldots, a_m(x) \in \mathbb{K}[[x]]$.

7.4 Closed Form Solutions

A natural tendency in the design of algorithms applicable to a specific class of sequences or power series is that for smaller input domains it is possible to solve harder problems by means of an efficient algorithm than for larger domains. If a sequence $(a_n)_{n=0}^{\infty}$ is given via a holonomic recurrence

$$p_0(n)a_n + p_1(n)a_{n+1} + \cdots + p_r(n)a_{n+r} = 0 \qquad (n \geq 0)$$

and a suitable number of initial values, we therefore might want to decide whether the sequence actually belongs to a smaller class: Is $(a_n)_{n=0}^{\infty}$ actually a polynomial sequence, or C-finite, or hypergeometric, or is its generating function algebraic? If so, we might want to compute appropriate alternative representations of $(a_n)_{n=0}^{\infty}$ so that algorithms which are only available in the smaller class become applicable.

Polynomial solutions of a holonomic recurrence equation can be found in very much the same way as described in Sect. 3.4 for a slightly different kind of equations: To determine a basis for the vector space of all polynomials $a(x) \in \mathbb{K}[x]$ such that

$$p_0(x)a(x) + p_1(x)a(x+1) + \cdots + p_r(x)a(x+r) = 0$$

first find a bound $d \in \mathbb{N}$ on the possible degree of $a(x)$, then make an ansatz $a(x) = a_0 + a_1 x + \cdots + a_d x^d$ with undetermined coefficients $a_0, \ldots, a_d$ and plug this ansatz into the equation. Finally compare coefficients of like powers of x to obtain a linear system of equations for the a_i. The solutions of this system are in one-to-one correspondence with the polynomial solutions of the recurrence.

The degree analysis from Sect. 3.4 can be generalized to holonomic equations of arbitrary order. Let

$$q_k(x) := \sum_{i=k}^{r} \binom{i}{k} p_i(x) \qquad (k = 0, \ldots, r)$$

and

$$\delta := \max_{0 \leq k \leq r} \left(\deg q_k(x) - k \right).$$

It can be shown (Problem 7.13) that if d is the degree of a polynomial solution $a(x)$, then $d \leq -\delta - 1$ or d is an integer root of the polynomial

$$\sum_{k=0}^{r} \left([x^{\delta+k}] q_k(x) \right) x^{\underline{k}} \in \mathbb{K}[x].$$

Therefore, one can choose as a degree bound simply the maximum between the nonnegative integer roots of this polynomial (if there are any) and $-\delta - 1$.

Using the algorithm for finding polynomial solutions as a subroutine, it is also possible to find hypergeometric solutions of a holonomic recurrence. This extension is known as Petkovšek's algorithm and works as follows. Suppose that there is a hypergeometric sequence $(h_n)_{n=0}^{\infty}$ which satisfies the recurrence

$$p_0(n)h_n + p_1(n)h_{n+1} + \cdots + p_r(n)h_{n+r} = 0 \qquad (n \geq 0).$$

We are only interested in sequences $(h_n)_{n=0}^{\infty}$ which are nonzero from some point on. According to the definition of hypergeometric sequences, there is some rational function $u(x) \in \mathbb{K}(x)$ with $h_{n+1} = u(n)h_n$ for all n where $u(n)$ is defined. Dividing the recurrence by h_n and rewriting the quotients h_{n+i}/h_n in terms of the rational function $u(x)$ gives the equation

$$p_0(x) + p_1(x)u(x) + p_2(x)u(x)u(x+1) + \cdots + p_r(x)\prod_{i=0}^{r-1} u(x+i) = 0.$$

Any rational solution $u(x)$ of this equation gives rise to a hypergeometric solution $(h_n)_{n=0}^{\infty}$ of the original equation, and vice versa. It remains to find the unknown rational function $u(x)$. If such a rational function exists, then it can be written in Gosper form, i.e., there will be polynomials $a(x), b(x), c(x) \in \mathbb{K}[x]$ with

$$u(x) = \frac{a(x+1)}{a(x)}\frac{b(x)}{c(x+1)} \quad \text{and} \quad \gcd(b(x), c(x+i)) = 1 \quad (i \in \mathbb{N} \setminus \{0\}).$$

Petkovšek pointed out that for every $u(x) \in \mathbb{K}(x)$, one can find $a(x), b(x), c(x) \in \mathbb{K}[x]$ which, in addition to Gosper's condition on the gcd of $b(x)$ and the shifts of $c(x)$, also satisfy $\gcd(a(x), b(x)) = \gcd(a(x), c(x)) = 1$. Furthermore one can require that $a(x)$ and $c(x)$ are monic, i.e., that $\mathrm{lc}_x a(x) = \mathrm{lc}_x c(x) = 1$. A Gosper form satisfying these additional requirements is called *Gosper-Petkovšek form*.

Suppose that $a(x), b(x), c(x)$ are a Gosper-Petkovšek form for the unknown rational function $u(x)$. Then the equation derived for $u(x)$ can be rewritten to

$$p_0(x) + p_1(x)\frac{a(x+1)}{a(x)}\frac{b(x)}{c(x+1)} + \cdots + p_r(x)\frac{a(x+r)}{a(x)}\prod_{i=0}^{r-1}\frac{b(x+i)}{c(x+i+1)} = 0.$$

Clearing denominators gives

$$\bar{p}_0(x)a(x) + \bar{p}_1(x)a(x+1) + \cdots + \bar{p}_r(x)a(x+r) = 0 \qquad \text{(P)}$$

where

$$\begin{aligned}
\bar{p}_0(x) &:= p_0(x)\, c(x+1)\, c(x+2)\, \cdots\, c(x+r), \\
\bar{p}_1(x) &:= p_1(x)\, b(x) \qquad c(x+2)\, \cdots\, c(x+r), \\
&\;\;\vdots \\
\bar{p}_r(x) &:= p_r(x)\, b(x) \qquad b(x+1)\, \cdots\, b(x+r-1).
\end{aligned}$$

Because of $b(x) \mid \bar{p}_i(x)$ for $i = 1,\ldots,r$ we must have $b(x) \mid \bar{p}_0(x)a(x)$, which, by the conditions of the Gosper-Petkovšek form, is only possible if $b(x) \mid p_0(x)$. Analogously, because of $c(x+r) \mid \bar{p}_i(x)$ for $i = 0,\ldots,r-1$ we must have $c(x+r) \mid \bar{p}_r(x)a(x+r)$, which, by the conditions of the Gosper-Petkovšek form, is only possible if $c(x) \mid p_r(x-r)$. Since both $p_0(x)$ and $p_r(x)$ are part of the input, we can simply try out all their divisors for $b(x)$ and $c(x)$, respectively, one after the other. There may be a lot of divisors, but up to a constant multiple, they are just finitely many.

So let $b_0(x), c_0(x)$ be a choice of monic divisors of $p_0(x)$ and $p_r(x-r)$, respectively. We next need to find multiplicative factors $z \in \mathbb{K}$ such that $b(x) = zb_0(x)$ and $c(x) = c_0(x)$ may possibly appear in the Gosper-Petkovšek form of a solution $u(x)$. This is not too hard a thing to do. In the end we want to solve (P) for polynomials $a(x)$, so all values of z which prevent this equation from having a polynomial solution are out of the business. To determine the values z which pass this filter, recall that $\deg a(x) = \deg a(x+i)$ for all $i \in \mathbb{N}$ and any polynomial $a(x)$, so that for $d := \max_{1 \le k \le r} \deg \bar{p}_k(x)$ we must have

$$[x^d](\bar{p}_0(x) + z\bar{p}_1(x) + \cdots + z^r \bar{p}_r(x)) = 0,$$

which is a polynomial equation for z with coefficients in $\mathbb{K}$. For all the (finitely many) values $z \in \mathbb{K}$ satisfying this condition, we finally determine the polynomial solutions $a(x)$ of (P). Each solution $a(x)$ gives rise to a hypergeometric solution $(h_n)_{n=0}^{\infty}$ with $h_{n+1} = z\frac{a(n+1)b_0(n)}{a(n)c_0(n+1)}h_n$ ($n \in \mathbb{N}$ sufficiently large) of the original equation.

In summary, Petkovšek's algorithm consists of the following steps.

1. For all pairs $(b_0(x), c_0(x))$ of monic polynomials with $b_0(x) \mid p_0(x)$ and $c_0(x) \mid p_r(x-r)$ perform the following steps 2–5.

2.   For $k = 0, \ldots, r$, let $\bar{p}_k(x) := p_k(x) \prod_{i=0}^{k-1} b_0(x+i) \prod_{i=k+1}^{r} c_0(x+i)$.

3.   Set $d := \max_{0 \le k \le r} \deg \bar{p}_k(x)$ and determine all $z \in \mathbb{K}$ such that

$$([x^d]\bar{p}_0(x)) + ([x^d]\bar{p}_1(x))z + ([x^d]\bar{p}_2(x))z^2 + \cdots + ([x^d]\bar{p}_r(x))z^r = 0.$$

   For all these $z \in \mathbb{K}$, perform the following steps 4–5.

4.   Find all $a(x) \in \mathbb{K}[x]$ such that

$$\bar{p}_0(x)a(x) + z\bar{p}_1(x)a(x+1) + \cdots + z^r\bar{p}_r(x)a(x+r) = 0.$$

5.   For each solution $a(x)$, output $u(x) := z\frac{a(x+1)b_0(x)}{a(x)c_0(x+1)}$.

Petkovšek's algorithm is complete in the sense that if for some equation it does not find a hypergeometric solution, then this proves that the equation has no hypergeometric solution.

To see the algorithm in action, let $(a_n)_{n=0}^{\infty}$ be the sequence defined by

$$2(n^2 + 2n + 2)a_n - (n^3 + 5n^2 + 4n + 4)a_{n+1} + (n^2 + 1)(n+2)a_{n+2} = 0 \qquad (n \ge 0)$$

and $a_0 = a_1 = 1$. We want to express a_n as a linear combination of hypergeometric sequences, if at all possible. All but two of the pairs

$$(b_0(x), c_0(x)) \in \{1, x^2 + 2x + 2\} \times \{1, x, (x-2)^2 + 1\}$$

to be considered lead in step 4 to equations which have no polynomial solution. The two interesting pairs are $(1, x)$ and $(1, 1)$. For the choice $b_0(x) = 1, c_0(x) = x$ we obtain

$$\bar{p}_0(x) = 2(x^2 + 2x + 2)(x+1)(x+2),$$
$$\bar{p}_1(x) = -(x^3 + 5x^2 + 4x + 4)(x+2),$$
$$\bar{p}_2(x) = (x^2 + 1)(x+2)$$

in step 2, and therefore the equation $2 - z + 0z^2 = 0$ in step 3. For its single solution $z = 2$, the equation in step 4 has the polynomial solution $a(x) = 1$. This solution gives rise to a hypergeometric solution $(h_n)_{n=0}^{\infty}$ with $h_{n+1} = \frac{2}{n+1} h_n$, e.g., $h_n = 2^n/n!$. The pair $(1, 1)$ yields the second solution $(n)_{n=0}^{\infty}$.

The solution space of the equation under consideration is two, and Petkovšek's algorithm has found the two linearly independent solutions $(2^n/n!)_{n=0}^{\infty}$ and $(n)_{n=0}^{\infty}$. Therefore it must be possible to express $(a_n)_{n=0}^{\infty}$ as a linear combination of these two solutions. Appropriate coefficients are found by matching initial values and solving a linear system. The final result is $a_n = 2^n/n! - n$ ($n \geq 0$).

In a bird's eye view, Petkovšek's algorithm may be described as applying suitable substitutions $a_n \mapsto r(n)a_n$ for some specific rational functions $r(x) \in \mathbb{K}(x)$ to a given holonomic recurrence equation to the end that polynomial solutions of the new equation correspond to hypergeometric solutions of the original one. There are algorithms based on similar techniques for finding, for instance, all the rational function solutions of a given holonomic differential equation. In combination with Theorems 7.1 and 4.3, such algorithms can be used to figure out whether a given holonomic sequence is actually C-finite. As the built-in solvers of most computer algebra systems are nowadays capable of finding rational solutions of differential equations, we omit a detailed description of these algorithms. For finding rational function solutions of recurrence equations, see Problem 7.17.

Finally, given a holonomic differential equation, it is also possible to determine its algebraic solutions. Algorithms for accomplishing this task depend heavily on the the theory of differential Galois theory [57], which is far beyond the scope of this book and mentioned here only for the sake of completeness.

7.5 The Tetrahedron for Holonomic Functions

One last time, let us go through the four vertices of the Concrete Tetrahedron: recurrence equations, generating functions, asymptotic estimates, and symbolic sums. Holonomic sequences are linked to linear recurrence equations by their very definition, and from Theorem 7.1 we know that their generating functions are precisely the holonomic power series.

A brute force approach to summation is that if $(a_n)_{n=0}^{\infty}$ is holonomic, then by Theorem 7.2, the sequence $(\sum_{k=0}^{n} a_k)_{n=0}^{\infty}$ is holonomic as well, and a recurrence for the sum can be obtained from a given recurrence for $(a_n)_{n=0}^{\infty}$. Besides this, it is sometimes possible to express symbolic sums involving a holonomic sequence in terms of the summand sequence. For example, if $(P_n(x))_{n=0}^{\infty}$ is the sequence of Legendre polynomials, then we have the summation identity

$$\sum_{k=0}^{n}(2k+1)P_k(x) = \frac{n+1}{1-x}P_n(x) - \frac{n+1}{1-x}P_{n+1}(x) \qquad (n \geq 0)$$

whose right hand side we may be willing to regard as a closed form for the sum on the left. Such closed form representations do not exist for every sum, but if they do, they may give more insight than a recursive description. So it is worthwhile to ask whether a given sum of a holonomic sequence has a closed form or not. The Abramov-van-Hoeij algorithm answers this question. Here is the precise problem statement. We are given a holonomic sequence satisfying

$$p_0(n)a_n + p_1(n)a_{n+1} + \cdots + p_r(n)a_{n+r} = 0 \qquad (n \geq 0),$$

where we will assume for simplicity that $p_r(x)$ has no nonnegative integer roots, and given a rational function $q(x) \in \mathbb{K}(x)$, and we want to find, if at all possible, rational functions $u_0(x), \ldots, u_{r-1}(x) \in \mathbb{K}(x)$ such that

$$(q(n)a_n)_{n=0}^{\infty} = \Delta(u_0(n)a_n + u_1(n)a_{n+1} + \cdots + u_{r-1}(n)a_{n+r-1})_{n=0}^{\infty}.$$

When this requirement is expanded into

$$u_{r-1}(n+1)a_{n+r} + (u_{r-2}(n+1) - u_{r-1}(n))a_{n+r-1}$$
$$+ (u_{r-3}(n+1) - u_{r-2}(n))a_{n+r-2}$$
$$+ \cdots$$
$$+ (u_0(n+1) - u_1(n))a_{n+1}$$
$$+ (-q(n) - u_0(n))a_n = 0 \qquad (n \geq 0),$$

it becomes tempting to simply match it to the given recurrence of $(a_n)_{n=0}^{\infty}$. From left to right, this suggests first $u_{r-1}(x) = p_r(x-1)$, then $u_{r-2}(x) = u_{r-1}(x-1) + p_{r-1}(x-1) = p_r(x-2) + p_{r-1}(x-1)$, and so on all the way down to $u_0(x)$, which is determined by comparing the coefficients of a_{n+1}. But also the coefficients of a_n have to match, so we must make sure that also the condition $-q(x) - u_0(x) = p_0(x)$ is satisfied. As we have to satisfy $r+1$ equations with only r unknowns, this might not work out. If it does, then we have found a solution to the summation problem. But if not, then it is still to early to conclude that no solution exists.

In general, it is necessary to adjust the recurrence appropriately before coefficient comparison yields a consistent solution. Following the advice of Abramov and van Hoeij, we introduce another unknown rational function, $m(x) \in \mathbb{K}(x)$, and multiply the recurrence of $(a_n)_{n=0}^{\infty}$ with this rational function prior to the coefficient comparison. This leads to $u_{r-1}(x) = m(x-1)p_r(x-1)$, $u_{r-2}(x) = m(x-2)p_r(x-2) + m(x-1)p_{r-1}(x-1)$, and so on all the way down to

$$u_0(x) = m(x-r)p_r(x-r) + \cdots + m(x-2)p_2(x-2) + m(x-1)p_1(x-1).$$

The final condition arising from comparing coefficients of a_n can be regarded as a recurrence equation for the as yet undetermined rational function $m(x)$. What we need is that $m(x) \in \mathbb{K}(x)$ satisfies the so-called *adjoint equation*

$$p_r(x)m(x) + p_{r-1}(x+1)m(x+1) + \cdots + p_0(x+r)m(x+r) = -q(x+r).$$

As we have all tools available to find rational function solutions to recurrence equations (Problem 7.17), we can algorithmically determine first $m(x)$ and then compute $u_{r-1}(x), \ldots, u_0(x)$ from this $m(x)$ and the coefficients of the given recurrence of $(a_n)_{n=0}^{\infty}$.

But what if no rational solution $m(x)$ exists? Could there be nevertheless some rational functions $u_0(x), \ldots, u_{r-1}(x)$ with

$$u_{r-1}(n+1)a_{n+r} + (u_{r-2}(n+1) - u_{r-1}(n))a_{n+r-1} + \cdots$$
$$\cdots + (u_0(n+1) - u_1(n))a_{n+1} + (-q(n) - u_0(n))a_n = 0 \quad (n \geq 0)?$$

If the defining recurrence

$$p_r(n)a_{n+r} + p_{r-1}(n)a_{n+r-1} + \cdots + p_0(n)a_n = 0 \quad (n \geq 0)$$

of $(a_n)_{n=0}^{\infty}$ is of minimal possible order r, then this cannot be, because then $m(x) := u_{r-1}(x+1)/p_r(x)$ is necessarily a rational solution of the adjoint equation. To see this, subtract the $m(n)$-fold of the defining recurrence from the relation for the assumed $u_0(x), \ldots, u_{r-1}(x)$ to obtain

$$\big(u_{r-2}(n+1) - u_{r-1}(n) - m(n)p_{r-1}(n)\big)a_{n+r-1}$$
$$+ \big(u_{r-3}(n+1) - u_{r-2}(n) - m(n)p_{r-2}(n)\big)a_{n+r-2}$$
$$+ \cdots$$
$$+ \big(u_0(n+1) - u_1(n) - m(n)p_1(n)\big)a_{n+1}$$
$$+ (-q(n) - u_0(n) - m(n)p_0(n))a_n = 0 \quad (n \geq 0).$$

Since r is minimal by assumption, all the coefficients in this relation must be identically zero. Being zero for all $n \in \mathbb{N}$ implies being zero as elements of $\mathbb{K}(x)$. Since shifting them does not make them nonzero, we will also have

$$u_{r-2}(x+2) - u_{r-1}(x+1) = m(x+1)p_{r-1}(x+1),$$
$$u_{r-3}(x+3) - u_{r-2}(x+2) = m(x+2)p_{r-2}(x+2),$$
$$\vdots$$
$$u_0(x+r) - u_1(x+r-1) = m(x+r-1)p_1(x+r-1),$$
$$-q(x+r) - u_0(x+r) = m(x+r)p_0(x+r).$$

If we now add up all these equations, there will be a telesoping effect on the left hand side which only leaves $-q(x+r)$ at the bottom and $u_{r-1}(x+1)$ at the top. Using also $u_{r-1}(x+1) = m(x)p_r(x)$ finally implies

$$-q(x+r) - p_r(x)m(x) = p_{r-1}(x+1)m(x+1) + \cdots + p_0(x+r)m(x+r).$$

This completes the argument that $m(x) = u_{r-1}(x+1)/p_r(x)$ is a solution of the adjoint equation. In conclusion, we have shown that the Abramov-van-Hoeij algorithm is a complete summation algorithm for holonomic sequences which are given in terms of a recurrence of minimal possible order.

We can use the algorithm to simplify, for example, the sum

$$\sum_{k=0}^{n} \frac{5k^2+1}{k+1} a_k,$$

where $(a_n)_{n=0}^{\infty}$ is defined by $a_0 = a_1 = 1$ and

$$(n+2)a_{n+2} - (2n-1)a_{n+1} + (n+2)a_n = 0 \qquad (n \geq 0).$$

The auxiliary equation

$$(x+2)m(x) - (2x+1)m(x+1) + (x+4)m(x+2) = -\frac{5(x+2)^2+1}{x+3}$$

happens to have the rational solution $m(x) = -\dfrac{5x^2+8x+2}{5(x+2)}$, so with

$$u_1(x) = (x+1)m(x-1) = \tfrac{1}{5}(1+2x-5x^2)$$

and $u_0(x) = xm(x-2) - (2x-3)m(x-1) = \dfrac{(x-1)(5x^2-7x+3)}{5(x+1)}$

we have

$$\frac{5n^2+1}{n+1}a_n = \left(u_1(n+1)a_{n+2} + u_0(n+1)a_{n+1}\right) - \left(u_1(n)a_{n+1} + u_0(n)a_n\right) \quad (n \geq 0).$$

Summing this equation finally gives the desired closed form:

$$\sum_{k=0}^{n} \frac{5k^2+1}{k+1}a_k = -\frac{5n^2+8n+2}{5}a_{n+2} + \frac{5n^3+3n^2+n}{5(n+2)}a_{n+1} + \frac{2}{5} \qquad (n \geq 0).$$

What remains to be discussed about the Tetrahedron for holonomic functions is the asymptotics. The reader who suspects that we will propose to determine the asymptotic behavior of a holonomic sequence by interpreting its generating function as an analytic function in a neighborhood of the origin, then expanding this function at the singularities of least absolute value in terms of the generalized series from Sect. 7.3, and then taking the growth of the series coefficients of the dominant term of this expansion, following step by step what we did in Chap. 6 for sequences with algebraic power series, this reader is not entirely mistaken. Indeed, this is what we propose to do. Singularities can only occur at points $\zeta \in \mathbb{C}$ which are roots of the polynomial coefficient $p_r(x)$ of the highest derivative in the differential equation for the generating function. Among the roots of this polynomial we determine the point closest to the origin and determine the generalized series solutions of the differential equation there. For translating these solutions into asymptotic information about the Taylor coefficients at the origin, we can exploit facts like

$$[x^n](1-x)^\alpha \sim \frac{1}{\Gamma(-\alpha)}n^{-\alpha-1} \qquad (n \to \infty, \; \alpha \in \mathbb{C} \setminus \mathbb{N}),$$

$$[x^n]\log(1-x)^m \sim \frac{(-1)^m}{n}\log(n)^{m-1} \qquad (n \to \infty, \; m \in \mathbb{N}),$$

$$\text{and } [x^n]\exp((1-x)^{-\alpha}) \sim e\frac{\mathbf{B}_n}{n!}\alpha^n \qquad (n \to \infty, \; \alpha > 0),$$

where $\mathbf{B}_n$ is the n-th Bell number.

However, the situation is in general not quite so simple. First, it can happen that a holonomic power series does not correspond to an analytic function in an open neighborhood of the origin, because the coefficients might grow too quickly for the series to converge. (Example: $(n!)_{n=0}^\infty$.) In this case, it sometimes helps to multiply the sequence under consideration by $(n!^{-m})_{n=0}^\infty$ for some suitably chosen $m \in \mathbb{N}$, then derive the asymptotics of the resulting sequence, and then multiply it by $n!^m$ to get the asymptotics of the original sequence.

The second potential complication is that a holonomic power series may correspond to an analytic function which has no singularities at all, because the sequence grows too slowly to cause a singularity at a finite distance of the origin. (Example: $(1/n!)_{n=0}^\infty$.) This case can sometimes be cured analogously to the former case, or other methods may become applicable which we do not describe here.

The third and most serious problem is to actually find the generalized series expansion that corresponds to a particular holonomic power series. For the case of algebraic functions we have seen in Chap. 6 that we must be careful to choose the branch which actually corresponds to the series under consideration. While this may at times be a technical business, it can always be done. After all, an algebraic curve can only give rise to finitely many branches. In contrast, the solutions of a holonomic differential equation form a vector space, so what previously was the problem of choosing the right branch out of a finite number of candidates now becomes the problem of choosing the coefficients of a linear combination of finitely many basis elements. More precisely, we have to do the following. We are given a holonomic power series $a(x) \in \mathbb{Q}[[x]]$ whose defining differential equation indicates a dominant singularity at $\zeta \in \mathbb{C}$. We determine the generalized series solutions of the differential equation with ζ as expansion point, call them $b_1(x), \ldots, b_r(x)$. For a suitably chosen open neighborhood U of 0 with ζ on its boundary, let $\bar{a}, \bar{b}_1, \ldots, \bar{b}_r : U \to \mathbb{C}$ denote the analytic functions corresponding to the formal objects $a(x), b_1(x), \ldots, b_r(x)$. Then the task is to determine the constants $c_1, \ldots, c_r \in \mathbb{C}$ with

$$\bar{a}(z) = c_1\bar{b}_1(z) + \cdots + c_r\bar{b}_r(z) \quad (z \in U).$$

At the time of writing, nobody knows how to do this. There are approaches which succeed in certain special situations, but general methods are still subject of ongoing research. With the currently available machinery it is often just possible to determine the asymptotic growth of a holonomic sequence up to a multiplicative constant which in general we do not know how to determine in closed form (and which may not even have a closed form in any reasonable sense of the word), but whose approximate value we can at any rate determine empirically to any desired accuracy.

7.6 Applications

Particular Permutations

There are $n!$ ways to permute the numbers $1, 2, \ldots, n$. But how many ways are there if we restrict the attention only to some particular type of permutations? Of course, this depends on the type, but often enough the counting sequence turns out to be holonomic. Here are some examples:

- The number of permutations with k cycles ($k \in \mathbb{N}$ fixed) is $(-1)^{n-k} S_1(n, k)$ where $S_1(n, k)$ are the Stirling numbers of first kind. The sequence $((-1)^{n-k} S_1(n, k))_{n=0}^{\infty}$ is holonomic for every fixed k (Problem 7.18).
- The number of permutations π where π^k is the identity ($k \in \mathbb{N}$ fixed) is given by $a_n^{(k)}$ where

$$\sum_{n=0}^{\infty} a_n^{(k)} \frac{x^n}{n!} = \exp\left(\sum_{d | k} \frac{x^d}{d} \right).$$

  For fixed k, the argument of the exponential on the right hand side is just a polynomial of degree k, so $(a_n^{(k)})_{n=0}^{\infty}$ is holonomic.
- An index x is called a *fixed point* of a permutation π if $\pi(x) = x$. The number of permutations with exactly k fixed points ($k \in \mathbb{N}$ fixed) is given by

$$T_{n,k} := \frac{n!}{k!} \sum_{j=0}^{n-k} \frac{(-1)^j}{j!} \qquad (n, k \geq 0).$$

  These numbers are known as *rencontre numbers*. Obviously they are holonomic for every fixed k.
- An index $x < n$ is called an *ascent* of a permutation π if $\pi(x) > \pi(x+1)$. The number of permutations with exactly k ascents ($k \in \mathbb{N}$ fixed) is given by

$$E_{n,k} := \sum_{j=0}^{k} (-1)^j \binom{n+1}{j} (k+1-j)^n \qquad (n, k \geq 0).$$

  These numbers are known as *Eulerian numbers*. Obviously they are holonomic (even C-finite) for every fixed k.

Many further kinds of permutations which are counted by holonomic sequences can be found in the literature.

High Performance Computations

The terms of a holonomic sequence $(a_n)_{n=0}^{\infty}$ can be computed efficiently by using the holonomic recurrence equation. This is exploited in high performance calculations for obtaining approximations of famous constants like π or e or γ to hundreds of millions of digits of accuracy. As a simple illustration, consider the continued fraction identity

$$\frac{4}{\pi} = 1 + \cfrac{1}{3 + \cfrac{4}{5 + \cfrac{9}{7 + \cfrac{16}{9 + \cdots}}}}.$$

Truncating the continued fraction at level 100, say, we expect to get a good rational approximations for the constant $4/\pi$. But bringing the truncated continued fraction into standard form p/q involves unpleasant calculations with rational numbers having long numerators and denominators if it is done naively. A more clever evaluation makes use of holonomic recurrence equations.

A general theorem about continued fractions says that if $(a_n)_{n=1}^{\infty}$ and $(b_n)_{n=1}^{\infty}$ are sequences of nonzero numbers, then

$$\cfrac{a_1}{b_1 + \cfrac{a_2}{b_2 + \cfrac{a_3}{\cdots + \cfrac{a_n}{b_n}}}} = \frac{p_n}{q_n} \qquad (n \geq 1)$$

where $(p_n)_{n=-1}^{\infty}$ and $(q_n)_{n=-1}^{\infty}$ are defined by

$$p_{n+2} = a_{n+2}p_{n+1} + b_{n+2}p_n \ (n \geq -1) \qquad p_{-1} = 1, p_0 = 0,$$
$$q_{n+2} = a_{n+2}q_{n+1} + b_{n+2}q_n \ (n \geq -1) \qquad q_{-1} = 0, q_0 = 1.$$

Note that if a_n and b_n are integers for all $n \in \mathbb{N}$ (as they are in the example continued fraction for $4/\pi$), then so are p_n and q_n ($n \in \mathbb{N}$). Note also that if $(a_n)_{n=1}^{\infty}$ and $(b_n)_{n=1}^{\infty}$ are polynomial sequences (as they are in the example above), then $(p_n)_{n=-1}^{\infty}$ and $(q_n)_{n=-1}^{\infty}$ are holonomic.

This can be used for computing numerator and denominator of the rational approximation of $4/\pi$ using arithmetic in $\mathbb{Z}$ rather than in $\mathbb{Q}$, which is significantly more efficient if numbers are long.

Bessel Functions

Bessel functions are, in a way, generalizations of the trigonometric functions $\sin(x)$ and $\cos(x)$. They are used in physics for describing the propagation of waves. The n-th Bessel function (of the first kind) is defined as the formal power series

$$J_n(x) := \sum_{k=0}^{\infty} \frac{(-1)^k}{k!(n+k)!} \left(\frac{x}{2}\right)^{n+2k} \in \mathbb{Q}[[x]].$$

It is not hard to see that for every fixed $n \in \mathbb{N}$ and any fixed number $z \in \mathbb{C}$ in place of x we obtain a convergent power series, so the formal power series corresponds to an analytic function defined on the entire complex plane.

Bessel functions are holonomic for every fixed $n \in \mathbb{N}$ by virtue of the differential equation

$$x^2 J_n''(x) + x J_n'(x) + (x^2 - n^2) J_n(x) = 0$$

which is easily deduced from the power series definition.

We can also keep x fixed (as a formal parameter) and let n run. This gives us a sequence $(J_n(x))_{n=0}^{\infty}$ in $\mathbb{Q}[[x]]$ which is holonomic as a sequence by virtue of the recurrence equation

$$J_{n+2}(x) - \frac{2(n+1)}{x} J_{n+1}(x) + J_n(x) = 0 \qquad (n \geq 0)$$

which is also easily deduced from the power series definition.

There are also mixed relations involving both derivation in x and shift in n at the same time; for example, we have

$$x J_n'(x) + x J_{n+1}(x) - n J_n(x) = 0 \qquad (n \geq 0),$$

which is also easily deduced from the power series definition.

There are Bessel functions of the second kind, two more kinds known as modified Bessel functions, and two more kinds known as spherical Bessel functions. Their definitions are similar to the definition of $J_n(x)$ and, in particular, they all can be seen as holonomic both as a sequence in n with x fixed or as a power series in x with n fixed.

Similar remarks apply to a plethora of other families of functions, including the classical families of orthogonal polynomials, which arise from all sorts of physical applications.

7.7 Problems

Problem 7.1 Prove that

$$H_n = \int_0^1 \frac{x^n - 1}{x - 1} dx \qquad (n \geq 0).$$

Problem 7.2 Simplify the following sums:

$$1. \ \sum_{k=1}^n kH_k, \qquad\qquad 2. \ \sum_{k=1}^n H_k^2, \qquad\qquad 3. \ \sum_{k=1}^n H_k^{(2)}.$$

Problem 7.3 Give an example of a sequence for each of the following types:

1. polynomial,
2. C-finite and hypergeometric but not polynomial,
3. C-finite but not hypergeometric,
4. hypergeometric but not C-finite,
5. algebraic and hypergeometric but not C-finite,
6. algebraic but neither hypergeometric nor C-finite,
7. hypergeometric but not algebraic,
8. holonomic but neither algebraic nor hypergeometric,
9. not holonomic.

Problem 7.4 Prove: if $(a_n)_{n=0}^\infty$ is holonomic and $(b_n)_{n=0}^\infty$ is such that $a_n \neq b_n$ for at most finitely many indices $n \in \mathbb{N}$, then $(b_n)_{n=0}^\infty$ is holonomic.

Problem 7.5 Prove part 2 of Theorem 7.1.

Problem 7.6 Prove part 4 of Theorem 7.2.

Problem 7.7 Which of the following power series in $\mathbb{Q}[[x]]$ are holonomic?

1. $\exp(x/\sqrt{1 - 4x})$,

2. $\dfrac{\exp(x)}{\sqrt{1 - 4x}}$,

3. $\exp(x) + \sum_{n=0}^\infty n! x^n$,

4. $\sum_{n=0}^\infty \left(\sum_{k=0}^n \binom{n}{k}^3 \right)^2 x^n$.

Problem 7.8 Prove that $\exp(e^x - 1)$ is not holonomic.

(*Hint:* Show that any potential holonomic differential equation for this series would give rise to an algebraic equation for $\exp(x)$.)

Problem 7.9 Let $s_n := \sum_{k=0}^{n}(-\frac{1}{2})^k \binom{n}{k}\binom{2k}{k}$. Theorem 5.1 implies that $(s_n)_{n=0}^{\infty}$ is holonomic. As we did not give a proof of this theorem, show by an independent argument using closure properties that $(s_n)_{n=0}^{\infty}$ is holonomic.

(*Hint:* It might be handy to go via the generating function and exploit the Euler transform from Problem 2.15.)

Problem 7.10 The sequence $(H_n(x))_{n=0}^{\infty}$ of Hermite polynomials is defined recursively by

$$H_{n+2}(x) - 2xH_{n+1}(x) + 2(n+1)H_n(x) = 0 \ (n \geq 0), \quad H_0(x) = 1, H_1(x) = 2x.$$

Prove the identity

$$\sum_{n=0}^{\infty} H_n(x)H_n(y)\frac{t^n}{n!} = \frac{1}{\sqrt{1-4t^2}}\exp\left(\frac{4t(xy - t(x^2 + y^2))}{1 - 4t^2}\right)$$

by making appropriate use of Theorem 7.2.

Problem 7.11 In $\mathbb{Q}(a,b,c)[[x]]$, prove Landen's hypergeometric transformation

$$_2F_1\left(\begin{array}{c}a,b\\2b\end{array}\bigg|\frac{4x}{(1+x)^2}\right) = (1+x)^{2a}{}_2F_1\left(\begin{array}{c}a,a-b+\frac{1}{2}\\b+\frac{1}{2}\end{array}\bigg|x^2\right).$$

Problem 7.12 Consider the differential equation

$$(5512 + 2809x + 52x^2)a(x) - 2(1378 + 1431x + 52x^2)a'(x) + (53x + 52x^2)a''(x) = 0.$$

As described in Sect. 7.3 one finds that it has two power series solutions of order $\alpha = 0$ and $\alpha = 53$, respectively. As also explained there, the 53rd coefficient of the power series corresponding to $\alpha = 0$ is undetermined. On the other hand, by Theorem 7.1 the coefficient sequence of this power series satisfies a holonomic recurrence equation of order at most four, and so we expect that the coefficient sequence is uniquely determined by four initial values. Why is this not a contradiction?

Problem 7.13 Check the claim made in Sect. 7.4 about the general degree bound for polynomials solutions of holonomic recurrence equations.

(*Hint:* Observe that the definition of $q_k(x)$ is made such that

$$p_0(x)a(x) + \cdots + p_r(x)a(x+r) = 0 \iff q_0(x)a(x) + \cdots + q_r(x)\Delta^r a(x) = 0,$$

and use $\Delta^r x^d = d^{\underline{r}}x^{d-r} + O(x^{d-r-1}).$)

Problem 7.14 Prove that the sum

$$\sum_{k=0}^{n} \binom{n}{k}^2 \binom{n+k}{k}^2$$

has no hypergeometric closed form by applying Petkovšek's algorithm to the recurrence equation found in Problem 5.15.

Problem 7.15 Consider the recurrence equation

$$(n+1)a_{n+2} - (n+2)n\,a_{n+1} - (n+2)(n+1)a_n = 0 \quad (n \geq 0).$$

Use Petkovšek's algorithm to find its hypergeometric solution $(h_n)_{n=0}^{\infty}$. There is a second solution $(u_n)_{n=0}^{\infty}$ which can be written as $u_n = h_n \sum_{k=0}^{n-1} \bar{h}_k$ where $(h_n)_{n=0}^{\infty}$ is the hypergeometric solution from before and $(\bar{h}_n)_{n=0}^{\infty}$ is some other hypergeometric sequence. Find $(\bar{h}_n)_{n=0}^{\infty}$.

Problem 7.16 Recall that the Legendre polynomials $P_n(x)$ satisfy

$$(n+2)P_{n+2}(x) - (3+2n)xP_{n+1}(x) + (n+1)P_n(x) = 0 \quad (n \geq 0)$$

and $P_0(x) = 1$, $P_1(x) = x$. Using Petkovšek's algorithm, determine all $\xi \in \mathbb{C}$ for which the sequence $(P_n(\xi))_{n=0}^{\infty}$ is hypergeometric.

Problem 7.17 This problem is about variations of Petkovšek's algorithm.

1. Let $p_0(x), \ldots, p_r(x) \in \mathbb{K}[x]$ and a hypergeometric sequence $(h_n)_{n=0}^{\infty}$ be given. Find a way to determine all hypergeometric solutions $(a_n)_{n=0}^{\infty}$ of the inhomogeneous recurrence equation

$$p_0(n)a_n + p_1(n)a_{n+1} + \cdots + p_r(n)a_{n+r} = h_n \quad (n \geq 0).$$

   (*Hint:* Make the equation homogeneous and then call Petkovšek's algorithm.)

2. Let $p_0(x), \ldots, p_r(x), q(x) \in \mathbb{K}[x]$ be given. Find a way to determine all rational functions $a(x) \in \mathbb{K}(x)$ such that

$$p_0(x)a(x) + p_1(x)a(x+1) + \cdots + p_r(x)a(x+r) = q(x).$$

   (*Hint:* Observe that every rational function is hypergeometric, and use the first part.)

Problem 7.18 For the Stirling numbers of the first kind, prove the identities

$$\begin{aligned}
S_1(n,1) &= (-1)^{n-1}(n-1)! & (n \geq 1), \\
S_1(n,2) &= (-1)^n(n-1)!H_{n-1} & (n \geq 1), \\
S_1(n,3) &= \tfrac{1}{2}(-1)^{n-1}(n-1)!(H_{n-1}^2 - H_{n-1}^{(2)}) & (n \geq 1),
\end{aligned}$$

and prove that for arbitrary fixed k, the sequence $(S_1(n,k))_{n=0}^{\infty}$ is holonomic.

Problem 7.19 The median of an array of numbers $(a_1, \ldots, a_n)$ is defined as the $\lceil n/2 \rceil$-smallest of these numbers. More generally, one might be interested in the i-th smallest number for a given i. One possibility to find this element in a given array of numbers is to first sort the array and then pick the i-th element.

Here is a better way: Choose a pivot element a_k at random, as in Quicksort, and separate, as in Quicksort, the elements smaller than a_k from those greater than a_k. After that, the position of a_k in the sorted array is known, say it is j. If $j = i$, then a_k is the i-th smallest element and we are done. If $j < i$, then the i-th smallest element of the whole array is the $(i - j)$-th smallest number in the part to the right of a_k, which we can find recursively without considering the part to the left of a_k any further. If $j > i$, then the i-th smallest element of the whole array is the i-th smallest number in the part to the left of a_k, which we can find recursively without considering the part to the right of a_k any further.

The cases $i = 1$ and $i = n$ may serve as recursion base, for in this case we just need to determine the minimum or maximum of the array, which can be done with only $n - 1$ comparisons.

1. Confirm that the proposed algorithm performs $\frac{1}{2}n(n-1)$ comparisons in the worst case.

2. Let $c_{n,i}$ be the number of comparisons the algorithm needs to determine the i-th smallest number in an array of n numbers. Derive a recurrence relation for the numbers $c_{n,i}$ from the description of the algorithm and suitable initial values by which the numbers $c_{n,i}$ can be computed.

3. Use the recurrence from part 2 to compute the numbers $c_{n, \lceil n/2 \rceil}$ for $n = 1, \ldots, 100$. (This should give something like $0, 1, \frac{8}{3}, \frac{53}{12}, \frac{197}{30}, \ldots$)

4. Use computer algebra to guess from these numbers a holonomic differential equation for the generating function $c(x) := \sum_{n=0}^{\infty} c_{n, \lceil n/2 \rceil} x^n$. (This should yield an equation of order 3 and degree 11.)

5. Use a computer algebra system to find three linearly independent closed form solutions $c_1(x), c_2(x), c_3(x)$ of the differential equation found in part 4.

6. Find constants $\alpha_1, \alpha_2, \alpha_3 \in \mathbb{Q}$ such that

$$c_{n, \lceil n/2 \rceil} = [x^n]\big(\alpha_1 c_1(x) + \alpha_2 c_1(x) + \alpha_3 c_3(x)\big)$$

for $n = 1, \ldots, 10$. (This ought to be possible.)

7. Determine the asymptotic behavior of

$$[x^n]\big(\alpha_1 c_1(x) + \alpha_2 c_1(x) + \alpha_3 c_3(x)\big)$$

as n goes to infinity.

8. Determine an explicit form for the coefficients

$$[x^n]\big(\alpha_1 c_1(x) + \alpha_2 c_1(x) + \alpha_3 c_3(x)\big)$$

in terms of harmonic numbers and/or other simple sums.

9. Conclude that – under the assumption that the guessed differential equation for $c(x)$ is correct – fast median search is more efficient than first sorting the whole array and then picking the middle element.

Conclude also – if applicable – that the tools presented in this book might be useful.

Appendix

A.1 Basic Notions and Notations

Domains

$\mathbb{N} = \{0,1,2,\dots\}$	natural numbers including zero
$\mathbb{Z}, \mathbb{Q}, \mathbb{R}, \mathbb{C}$	integers, rational, real, and complex numbers
$\mathbb{K}$	an arbitrary field of characteristic zero
$\bar{\mathbb{K}}$	the algebraic closure of $\mathbb{K}$
$\mathbb{K}^n$	vector space over $\mathbb{K}$ of dimension n
$\mathbb{K}^{n \times m}$	vector space of $n \times m$ matrices over $\mathbb{K}$
R	an arbitrary commutative ring containing $\mathbb{Q}$ as subring
$R[x]$	polynomials in x with coefficients in R (p. 44)
$\mathbb{K}(x)$	rational functions in x with coefficients in $\mathbb{K}$ (p. 44)
$R[[x]]$	formal power series in x with coefficients in R (p. 18)
$\mathbb{K}((x))$	formal Laurent series in x with coefficients in $\mathbb{K}$ (p. 23)
$R^{\mathbb{N}}$	ring of sequences over R

Logic

$A \overset{!}{=} B$	desired equality: A shall be (made) equal to B
$A \overset{?}{=} B$	questioned equality: Are A and B equal?

Asymptotics

$\lim_{n \to \infty} a_n = c$	limit of a convergent sequence: $\forall\, \varepsilon > 0\ \exists\, n_0\ \forall\, n \geq n_0 : $ $\|a_n - c\| < \varepsilon$
$a_n \sim b_n\ (n \to \infty)$	asymptotic equivalence of sequences: $\lim_{n \to \infty} \frac{a_n}{b_n} = 1$

M. Kauers, P. Paule, *The Concrete Tetrahedron*
© Springer-Verlag/Wien 2011

$\lim_{z \to \zeta} a(z) = c$ limit of functions: $\forall\, \varepsilon > 0\ \exists\, \delta > 0\ \forall\, z : |z - \zeta| < \delta\ \Rightarrow$ $|a(z) - c| < \varepsilon$

$a(z) \sim b(z)\ (z \to \zeta)$ asymptotic equivalence of functions: $\lim_{z \to \zeta} \frac{a(z)}{b(z)} = 1$

$a_n = O(b_n)\ (n \to \infty)$ O-notation: $\exists\, c \in \mathbb{R}\ \exists\, n_0\ \forall\, n \geq n_0 : |a_n| \leq c b_n$

$a_n = o(b_n)\ (n \to \infty)$ o-notation: $\lim_{n \to \infty} \frac{a_n}{b_n} = 0$

Sequences

$(a_n)_{n=0}^{\infty}$ the sequence $a_0, a_1, a_2, \ldots$; formally $a \colon \mathbb{N} \to \mathbb{K},\ n \mapsto a_n$

$(a_{n,k})_{n,k=0}^{\infty}$ notation for a bivariate sequence

Δa_n forward difference: $a_{n+1} - a_n$

$\sum_{k=u}^{v} a_k$ the sum $a_u + a_{u+1} + \cdots + a_v$. If $v < u$, the sum is defined as 0.

$\prod_{k=u}^{v} a_k$ the product $a_u a_{u+1} \cdots a_v$. If $v < u$, the product is defined as 1.

$p(x) \bullet (a_n)_{n=0}^{\infty}$ action of the polynomial $p(x) \in \mathbb{K}[x]$ on the sequence $(a_n)_{n=0}^{\infty}$ (p. 68)

Linear Algebra

$(a_i)_{i=1}^{n}$ the vector $(a_1, a_2, \ldots, a_n) \in \mathbb{K}^n$

$((a_{i,j}))_{i,j=1}^{n}$ the $n \times n$ matrix with entry $a_{i,j}$ in row i and column j.

$\det A$ determinant of the matrix $A \in \mathbb{K}^{n \times n}$

$\dim V$ dimension of the vector space V

$V \oplus W$ direct sum of the vector spaces V and W

Polynomials and formal power series

$[x^n] a(x)$ coefficient of the monomial x^n in $a(x)$

$\deg_x p(x),\ \deg p(x)$ degree of the polynomial $p(x)$: $\max\{n \in \mathbb{N} : [x^n] p(x) \neq 0\}$

$\mathrm{lc}_x p(x),\ \mathrm{lc}\, p(x)$ leading coefficient of the polynomial $p(x)$: $\mathrm{lc}\, p(x) = [x^{\deg p(x)}] p(x)$

$\mathrm{res}_x(p(x), q(x))$ resultant of the polynomials $p(x)$ and $q(x)$ (p. 124)

$\gcd(p(x), q(x))$ greatest common divisor of the polynomials $p(x)$ and $q(x)$ (p. 168)

$\mathrm{ord}\, a(x)$ order of the series $a(x)$ (p. 23)

$a(x) \odot b(x)$ Hadamard product of $a(x)$ and $b(x)$ (p. 18)

$D_x a(x)$ (formal) derivative of $a(x)$ (p. 20)

$\int_x a(x)$ (formal) integral of $a(x)$ (p. 20)

$\lim_{k \to \infty} a_k(x)$ limit of a sequence of power series (p. 24)

$a(b(x))$ composition of $a(x)$ with $b(x)$ (p. 25)

$\sum_{k=0}^{\infty} a_k(x),\ \prod_{k=0}^{\infty} a_k(x)$ infinite sum and product of a sequence of power series (p. 28)

Common particular functions and sequences

$\lfloor x \rfloor$	floor of x: greatest $n \in \mathbb{Z}$ with $n \leq x$
$\lceil x \rceil$	ceiling of x: smallest $n \in \mathbb{Z}$ with $n \geq x$
$x^{\overline{n}}$	rising factorial: $x^{\overline{n}} = x(x+1)\cdots(x+n-1)$
$x^{\underline{n}}$	falling factorial: $x^{\underline{n}} = x(x-1)\cdots(x-n+1)$
$n!$	factorial: $n! = n^{\underline{n}} = 1^{\overline{n}} = 1\cdot 2\cdot 3\cdots n$
$\binom{x}{k}$	binomial coefficient (p. 88)
$S_1(n,k), S_2(n,k)$	Stirling numbers of the first and second kind, respectively (p. 45)
B_n	Bernoulli numbers (p. 23)
$\mathbf{B}_n$	Bell numbers (p. 26)
C_n	Catalan numbers (p. 113)
p_n	partition numbers (p. 27)
F_n	Fibonacci numbers (p. 63)
H_n	harmonic numbers (p. 8)
$H_n^{(r_1,\ldots,r_m)}$	generalized harmonic numbers (p. 138)
$\Gamma(z)$	gamma function (p. 91)
$T_n(x), P_n(x), H_n(x)$	Chebyshev polynomials of the first kind (p. 54), Legendre polynomials (p. 132), Hermite polynomials (p. 161)
$J_n(x)$	Bessel functions (p. 159)

A.2 Basic Facts from Computer Algebra

Classical computer algebra focusses on algorithms for computing with polynomials. We list here only those very basic results from computer algebra which are used in some part of this book. In particular, we only state here *what* can be computed and not *how* to carry out these computations. For further background and many additional algorithmic results about polynomials, we refer to the standard text books on computer algebra [22, 63, 58].

In order to do computations in $\mathbb{K}[x]$, some technical conditions have to be imposed on the field $\mathbb{K}$. For example, there is no way to do exact computations in $\mathbb{R}$ or $\mathbb{C}$, because the elements of these fields are inherently infinite objects and cannot be faithfully represented in a (finite) computer. But computations in the field $\mathbb{Q}$ of rational numbers, in algebraic number fields such as $\mathbb{Q}(\sqrt{2})$, in rational function fields $\mathbb{Q}(x)$, or in finite fields $\mathbb{Z}_p$ can be carried out exactly. Such fields are called admissible.

Let $\mathbb{K}$ be an admissible field. Then there are algorithms

- to compute for given $a(x), b(x) \in \mathbb{K}[x]$ the coefficients of $a(x)+b(x)$, $a(x)b(x)$, $a(b(x))$, etc.;
- to compute for given $a(x), b(x) \in \mathbb{K}[x]$ the unique polynomials $q(x), r(x) \in \mathbb{K}[x]$ with $r(x) = 0$ or $\deg r(x) < \deg b(x)$ such that $a(x) = q(x)b(x) + r(x)$ (*division with remainder*);

- to compute for given $a(x), b(x) \in \mathbb{K}[x]$ the unique monic polynomial $g(x) :=$ $\gcd_x(a(x), b(x)) := \gcd(a(x), b(x)) \in \mathbb{K}[x]$ of maximal degree with $g(x) \mid a(x)$ and $g(x) \mid b(x)$ (*greatest common divisor*);

- to compute for given $a(x), b(x) \in \mathbb{K}[x]$ the unique polynomials $u(x), v(x) \in \mathbb{K}[x]$ with $g(x) := \gcd(a(x), b(x)) = u(x)a(x) + v(x)b(x)$ and $\deg u(x) < \deg b(x) - \deg g(x)$ and $\deg v(x) < \deg a(x) - \deg g(x)$ (*Bezout coefficients*);

- to compute for a given set $\{(x_0, y_0), \ldots, (x_n, y_n)\} \subseteq \mathbb{K}^2$ with $x_i \neq x_j$ $(i \neq j)$ the unique polynomial $p(x) \in \mathbb{K}[x]$ of degree at most n such that $p(x_i) = y_i$ $(i = 0, \ldots, n)$ (*polynomial interpolation*);

- to compute for a given $a(x) \in \mathbb{K}[x]$ a polynomial $\bar{a}(x) \in \mathbb{K}[x]$ which has the same roots as $a(x)$ in $\bar{\mathbb{K}}$ but not repeated factors (*square free part*);

- to compute for a given $a(x) \in \mathbb{K}[x]$ a representation $a(x) = a_1(x)a_2(x)^2 \cdots a_m(x)^m$ where $a_1(x), \ldots, a_m(x)$ have no repeated factors and mutually disjoint sets of roots (*square free decomposition*).

Under slightly more restrictive assumptions on $\mathbb{K}$ (which are still satisfied for all fields appearing in this book), there are also algorithms

- to determine for a given polynomial $a(x) \in \mathbb{K}[x]$ the set of all $n \in \mathbb{Z}$ such that $a(n) = 0$; (*integer roots*),

- to decide for a given polynomial $a(x) \in \mathbb{K}[x]$ whether there exists $b(x) \in \mathbb{K}[x]$ with $1 < \deg b(x) < \deg a(x)$ and $b(x) \mid a(x)$ (*irreducibility*);

- to compute for a given polynomial $a(x) \in \mathbb{K}[x]$ the unique factorization $a(x) = c p_1(x)^{e_1} \cdots p_m(x)^{e_m}$ into monic irreducible polynomials (*factorization*).

All these operations can be performed efficiently, which in theory means the algorithms performing the listed operations have a runtime complexity which depends polynomially on the size of the input. In practice, using the most careful implementations of the most advanced algorithms on the most recent hardware, the manipulation of polynomials with some ten thousand terms may well be feasible.

Some of the items in the list above generalize from univariate polynomials to polynomials in several variables. In particular it is possible to compute the greatest common divisor of multivariate polynomials. However, for multivariate polynomials there is in general no Bezout representation of the greatest common divisor, because $\mathbb{K}[x_1, \ldots, x_n]$ is not a Euclidean domain. It is a unique factorization domain, though, and for many types of fields $\mathbb{K}$ there are algorithms that split a given multivariate polynomial into irreducible factors.

A.3 A Collection of Formal Power Series Identities

We list here the most frequently needed identities related to quantities appearing in the text. Additional series expansions can be found in the book of Wilf [62]. Further identities for special functions can be found in tables like the classical volume of Abramowitz and Stegun [5] or its recent successor project [19]. A version of the latter is also available electronically at http://dlmf.nist.gov (Digital Library of Mathematical Functions)

$$\log(1-x) = \sum_{n=1}^{\infty} \frac{1}{n} x^n$$

$$\sum_{n=0}^{\infty} x^n = \frac{1}{1-x}$$

$$\exp(x) = \sum_{n=0}^{\infty} \frac{1}{n!} x^n$$

$$\sum_{n=0}^{\infty} n^{\overline{k}} x^n = \frac{k!}{(1-x)^{k+1}}$$

$$\arctan(x) = \sum_{n=0}^{\infty} \frac{(-1)^n}{2n+1} x^{2n+1}$$

$$\sum_{n=0}^{\infty} H_n x^n = \frac{1}{1-x} \log \frac{1}{1-x}$$

$$\sin(x) = \sum_{n=0}^{\infty} \frac{(-1)^n}{(2n+1)!} x^{2n+1}$$

$$\sum_{n,k=0}^{\infty} \binom{n}{k} x^n y^k = \frac{1}{1-x-xy}$$

$$\cos(x) = \sum_{n=0}^{\infty} \frac{(-1)^n}{(2n)!} x^{2n}$$

$$\sum_{n=0}^{\infty} \binom{n}{k} x^n = \frac{x^k}{(1-x)^{k+1}}$$

$$(1+x)^\lambda = \sum_{n=0}^{\infty} \binom{\lambda}{n} x^n$$

$$\sum_{n=0}^{\infty} C_n x^n = \frac{1-\sqrt{1-4x}}{2x}$$

$$x^n = \sum_{k=0}^{n} S_2(n,k) x^{\underline{k}}$$

$$\sum_{k=0}^{n} S_1(n,k) x^k = x^{\underline{n}}$$

$$\frac{x}{\exp(x)-1} = \sum_{n=0}^{\infty} \frac{B_n}{n!} x^n \ \text{(Bernoulli numbers)}$$

$$\sum_{n=0}^{\infty} F_n x^n = \frac{x}{1-x-x^2}$$

$$\exp(e^x - 1) = \sum_{n=0}^{\infty} \frac{\mathbf{B}_n}{n!} x^n \ \text{(Bell numbers)}$$

$$\sum_{n=0}^{\infty} T_n(x) y^n = \frac{1-xy}{1-2xy+y^2}$$

$$P_n(x) = \sum_{k=0}^{n} \frac{(-1)^k}{2^n} \binom{n}{k} \binom{2n-2k}{k} x^{n-2k}$$

$$\sum_{n=0}^{\infty} P_n(x) y^n = \frac{1}{\sqrt{1-2xy+y^2}}$$

$$J_n(x) = \sum_{k=0}^{\infty} \frac{(-1)^k}{k!(n+k)!} \left(\frac{x}{2}\right)^{n+2k}$$

$$\sum_{n=0}^{\infty} \frac{H_n(x)}{n!} y^n = \exp(2xy - y^2)$$

$${}_pF_q \left(\begin{matrix} a_1, a_2, \ldots, a_p \\ b_1, \ldots, b_q \end{matrix} \bigg| x \right) = \sum_{k=0}^{\infty} \frac{a_1^{\overline{k}} a_2^{\overline{k}} \cdots a_p^{\overline{k}}}{b_1^{\overline{k}} \cdots b_q^{\overline{k}}} \frac{x^k}{k!}$$

A.4 Closure Properties at One Glance

We summarize the closure properties of the classes of formal power series discussed in this book. Not all of the facts stated in the tables below are explicitly mentioned in the text. The following abbreviations are used:

F   polynomial generating function / finite sequence
P   generating function of the form $p(x)/(1-x)^d$ /
    sequences which from some index on agree with
    a polynomial sequence
C   rational generating function / C-finite sequence
Hg hypergeometric series
A   algebraic series
H   holonomic series
S   formal power series

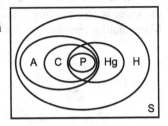

Unary operations

For $1/a(x)$ it is assumed that $a(0) \neq 0$.

$a(x)$	$\frac{1}{a(x)}$	$\frac{a(x)-a(0)}{x}$	$\frac{1}{1-x}a(x)$	$D_x a(x)$	$\int_x a(x)$
F	C	F	P	F	F
P	C	P	P	P	H
C	C	C	C	C	H
Hg	S	Hg	H	Hg	H
A	A	A	A	A	H
H	S	H	H	H	H
S	S	S	S	S	S

Binary operations

For $a(b(x))$ it is assumed that $b(0) = 0$.

$a(x)$	$b(x)$	$a(x)+b(x)$	$a(x)b(x)$	$a(x)\odot b(x)$	$a(b(x))$
F	F	F	F	F	F
F	P	P	P	F	P
P	F	P	P	F	C
P	P	P	P	P	C
C	C	C	C	C	C
Hg	Hg	H	H	Hg	S
A	A	A	A	H	A
H	F	H	H	F	H
H	A	H	H	H	H
H	H	H	H	H	S
S	S	S	S	S	S

A.5 Software

The algorithms described in this book as well as many other algorithms which are useful for solving problems related to the Concrete Tetrahedron have been implemented and are available in computer algebra systems. We give here a brief overview of some of the most basic maneuvers in Maple and Mathematica. Details can be found in the documentation of the respective pieces of software.

Maple

Gosper's and Zeilberger's algorithm are part of the built-in *sum* command:

$sum(k\,k!, k = 0..n);$

$$(n+1)! - 1 \tag{1}$$

$sum(\text{binomial}(n,k), k = 0..n);$

$$2^n \tag{2}$$

For definite sums, the *sum* command applies first Zeilberger's algorithm to find a recurrence for the sum and then uses Petkovšek's algorithm to solve this recurrence. If no closed form exists, some standardized format of the sum is returned:

$sum(\text{binomial}(n,k)^2 \text{binomial}(n+k,k)^2, k = 0..n);$

$$\text{hypergeom}([-n, -n, n+1, n+1], [1,1,1], 1) \tag{3}$$

For getting the recurrence of a sum, there is the command *Zeilberger* in the *Hypergeometric* section of the package *SumTools*:

$SumTools[Hypergeometric][Zeilberger](\text{binomial}(n,k)^2 \text{binomial}(n+k,k)^2, n, k, N);$

$$\left[(6n^2 + 12n + 8 + n^3)N^2 + (-34n^3 - 153n^2 - 231n - 117)N + 3n + 1 + n^3 + 3n^2, \tag{4} \right.$$
$$\left. \frac{(-4 - 2n^2 - 6n - \frac{3}{2}k + k^2)k^4 \text{binomial}(n,k)^2 \text{binomial}(n+k,k)^2 (16n+24)}{(-n-1+k)^2 (-n-2+k)^2} \right]$$

Hypergeometric solutions of holonomic recurrence equations can be found via the built-in *rsolve* command.

$rsolve(\{2(n^2 + 2n + 2)a(n) - (n^3 + 5n^2 + 4n + 4)a(n+1) + (n^2 + 1)(n+2)a(n+2) = 0, a(0) = 1, a(1) = 1\}, a(n));$

$$-n + \frac{2^n}{\Gamma(n+1)} \tag{5}$$

A description of Maple's procedures for hypergeometric summation is given in [3].

Functions for executing closure properties for holonomic sequences and power series are available in the package *gfun* by Salvy and Zimmermann [49]. This package allows, for example, to compute a recurrence for $(a_n + b_n)_{n=0}^{\infty}$ or $(a_n b_n)_{n=0}^{\infty}$ given recurrences for $(a_n)_{n=0}^{\infty}$ and $(b_n)_{n=0}^{\infty}$. Note that both input recurrences must be represented with the same function symbol, and that this symbol is also used in the output recurrence:

$gfun[`rec+rec`](\{f(n+2)=f(n+1)+f(n),f(0)=0,f(1)=1\},$

$\{f(n+1)=\frac{n+1}{n+2}f(n),f(0)=1\},f(n))$

$\{(n^3+9n^2+22n+14)f(n)+(-n^2-9n-14)f(n+1)+(-19n^2-53n$ (6)

$-42-2n^3)f(n+2)+(n^3+10n^2+31n+28)f(n+3),f(0)=1,f(1)=\frac{3}{2},f(2)=\frac{4}{3}\}$

$gfun[`rec*rec`](\{f(n+2)=f(n+1)+f(n),f(0)=0,f(1)=1\},$

$\{f(n+1)=\frac{n+1}{n+2}f(n),f(0)=1\},f(n))$

$\{(-n-1)f(n)+(-n-2)f(n+1)+(n+3)f(n+2),f(0)=0,f(1)=\frac{1}{2}\}$ (7)

The package also provides guessing facilities:

$gfun[listtorec]([1,1,2,5,14,42,132,429,1430,4862,16796],f(n));$

$[\{(-4n-2)f(n)+(n+2)f(n+1),f(0)=1\},ogf]$ (8)

$gfun[listtoalgeq]([1,1,2,5,14,42,132,429,1430,4862,16796],f(x));$

$[-1+f(x)-xf(x)^2,ogf]$ (9)

Conversely, the *series* command determines the first terms of a series

$series((1-sqrt(1-4x))/(2x),x=0,10);$

$1+x+2x^2+5x^3+14x^4+42x^5+132x^6+429x^7+1430x^8+O(x^9)$ (10)

Also Puiseux expansions can be computed with this command, for instance the expansion of the generating function for Catalan numbers at its singularity $x=1/4$:

$series((1-sqrt(1-4x))/(2x),x=1/4,3);$

$4-4\mathrm{i}(x-\frac{1}{4})^{1/2}-8x+16\mathrm{i}(x-\frac{1}{4})^{3/2}+32(x-\frac{1}{4})^2-64\mathrm{i}(x-\frac{1}{4})^{5/2}+O((x-\frac{1}{4})^3)$ (11)

Mathematica

Like its cousin in Maple, the **Sum** command in Mathematica resorts to the algorithms of Gosper and Zeilberger for simplifying hypergeometric sums. As an alternative to this built-in command, there are also special purpose add-on packages which provide functionality for doing summation and related calculations. Some packages are available at

http://www.risc.jku.at/research/combinat/software/

There is a package of Paule and Schorn [42] for hypergeometric summation, a package by Mallinger [39] for holonomic closure properties and guessing, and several other useful packages with sophisticated algorithms which are not discussed in this book, such as Schneider's package Sigma for simplification of nested sums [50], Kauers's multivariate guessing library [31], Wegschaider's package for multivariate hypergeometric summation [61], or Koutschan's package for multivariate holonomic series [36]. Of these, we give here just some examples for the Paule-Schorn package and the Mallinger package:

In[1]:= **≪ zb.m**

Fast Zeilberger Package by Peter Paule and Markus Schorn (enhanced by Axel Riese – © RISC Linz – V 3.54 (02/23/05)

In[2]:= **Gosper[$k\,k!$, k]**

Out[2]= $\{k\,k! = \Delta_k k!\}$

In[3]:= **Gosper[$k\,k!$, $\{k, 0, n\}$]**

If `n' is a natural number, then:

Out[3]= $\{\sum_{k=0}^{n} k\,k! == -1 + (n+1)n!\}$

In[4]:= **Zb[Binomial[n,k]^2 Binomial[$n+k,k$]^2, k, n]**

If `n' is a natural number, then:

Out[4]= $\{(1+n)^2 F[k,n] - (3+2n)(39+51n+17n^2)F[k,n+1] + (2+n)^3 F[k,n+2]$
$== \Delta_k F[k,n]R[k,n]\}$

In[5]:= **Show[F]**

Out[5]= $\binom{n}{k}^2 \binom{k+n}{n}^2$

In[6]:= **Show[R]**

Out[6]= $\dfrac{k^4(-12k(3+2n)+8k^2(3+2n)-16(3+2n)(2+3n+n^2))}{(1-k+n)^2(2-k+n)^2}$

In[7]:= **Zb[Binomial[n,k]^2 Binomial[$n+k,k$]^2, $\{k,0,n\}$, n]**

If `n' is a natural number, then:

Out[7]= $\{(1+n)^2 \text{SUM}[n] - (3+2n)(39+51n+17n^2)\text{SUM}[n+1] + (2+n)^3\text{SUM}[n+2] == 0\}$

In[8]:= **≪ GeneratingFunctions.m**

GeneratingFunctions Package by Christian Mallinger – © RISC Linz – V 0.68 (07/17/03)

In[9]:= **REPlus[$\{f[n+2] == f[n] + f[n+1], f[0] == 0, f[1] == 1\}$, $\{f[n+1] == \frac{n+1}{n+2}f[n], f[0] == 1\}$, $f[n]$]**

Out[9]= $\{(1+n)(14+8n+n^2)f[n] - (2+n)(7+n)f[1+n] - (3+n)(14+13n+2n^2)f[2+n] + (4+$
$n)(7+6n+n^2)f[3+n] == 0, f[0] == 1, f[1] == \frac{3}{2}, f[2] == \frac{4}{3}\}$

In[10]:= **REHadamard[$\{f[n+2] == f[n] + f[n+1], f[0] == 0, f[1] == 1\}$, $\{f[n+1] == \frac{n+1}{n+2}f[n], f[0] == 1\}$, $f[n]$]**

Out[10]= $\{(-1-n)f[n] + (-2-n)f[1+n] + (3+n)f[2+n] == 0, f[0] == 0, f[1] == \frac{1}{2}\}$

In[11]:= **GuessRE[$\{1,1,2,5,14,42,132,429,1430,4862,16796\}$, $f[n]$]**

Out[11]= $\{\{-2(1+2n)f[n] + (2+n)f[1+n] == 0, f[0] == 1\}, \text{ogf}\}$

Mathematica's built-in **Series** command is useful for going the other direction:

In[12]:= **Series[$(1-\text{Sqrt}[1-4x])/(2x)$, $\{x,0,5\}$]**

Out[12]= $1 + x + 2x^2 + 5x^3 + 14x^4 + 42x^5 + O[x]^6$

In[13]:= **Series[$(1-\text{Sqrt}[1-4x])/(2x)$, $\{x,1/4,3\}$]**

Out[13]= $2 - (4i)\sqrt{-\frac{1}{4}+x} - 8(-\frac{1}{4}+x) + (16i)(-\frac{1}{4}+x)^{3/2} + 32(-\frac{1}{4}+x)^2 - (64i)(-\frac{1}{4}+x)^{5/2} -$
$128(-\frac{1}{4}+x)^3 + O[-\frac{1}{4}+x]^{7/2}$

A Mathematica implementation of Petkovšek's algorithm by Petkovšek himself is available at

http://www.fmf.uni-lj.si/~petkovsek/software.html

This implementation returns the shift quotients of hypergeometric solutions as output:

In[14]:= ≪ **Hyper.m**
In[15]:= **Hyper**[$2(n^2 + 2n + 2)a[n] - (n^3 + 5n^2 + 4n + 4)a[n+1] + (n^2 + 1)(n+2)a[n+2], a[n]$,
           Solutions → **All**]

Out[15]= $\left\{ \dfrac{2}{n+1}, \dfrac{1+n}{n} \right\}$

A.6 Solutions to Selected Problems

Problem 1.1   Induction on n.

Problem 1.2   1. $(n+1)^2$; 2. $\frac{1}{6}n(n+1)(2n+1)$; 3. $\frac{1}{4}n^2(n+1)^2$.

Problem 1.3 $\frac{n+1}{2n}$.

Problem 1.4   From the definition of the Riemann integral we obtain

$$\log(n) = \int_1^n \frac{1}{x}\,dx \le \sum_{k=1}^{n-1}{}' 1 \times \max_{x \in [k,k+1]} \frac{1}{x} = \sum_{k=1}^{n-1} \frac{1}{k} \le H_n,$$

so $H_n - \log(n) \ge 0$ $(n \ge 1)$. Secondly, from

$$\left(1 + \frac{1}{n}\right)^{n+1} \ge e \qquad (n \ge 1)$$

we obtain

$$(n+1)\big(\log(n+1) - \log(n)\big) \ge 1 \qquad (n \ge 1)$$

by taking logarithm on both sides. Dividing both sides by $n+1$ and using $H_{n+1} - H_n = \frac{1}{n+1}$ gives

$$H_n - \log(n) \ge H_{n+1} - \log(n+1) \qquad (n \ge 1),$$

so $H_n - \log(n)$ is decreasing. The claim follows.

Problem 1.5   1. $\frac{x}{(1-x)^2}$; 2. $-\frac{x(x+1)}{(1-x)^3}$; 3. $-\frac{1}{x}\log(1-x)$.

Problem 1.6 $a(2x)$.

Problem 1.7   Use the relation $\log(-z) = i\pi + \log(z)$.

Problem 2.2   Use the order as degree function. Then the greatest common divisor of two power series $a(x), b(x) \in \mathbb{K}[[x]]$ is the series of lower order, for whenever $\operatorname{ord} a(x) < \operatorname{ord} b(x)$, then $a(x) \mid b(x)$.

Problem 2.3   Using the definition, the Cauchy product formula, and the binomial theorem, we can calculate

$$
\exp(ax)\exp(bx) = \left(\sum_{n=0}^{\infty} a^n \frac{x^n}{n!} \right) \left(\sum_{n=0}^{\infty} b^n \frac{x^n}{n!} \right) = \sum_{n=0}^{\infty} \left(\sum_{k=0}^{n} \binom{n}{k} a^k b^{n-k} \right) \frac{x^n}{n!}
$$

$$
= \sum_{n=0}^{\infty} (a+b)^n \frac{x^n}{n!} = \exp((a+b)x).
$$

Problem 2.4   Hopefully none.

Problem 2.7   If $D(a) = 0$ and $D(b) = 0$, then $D(a+b) = D(a) + D(b) = 0$ and $D(ab) = D(a)b + aD(b) = 0$.

Problem 2.8   1. $(n+1)2^n$; 2. $\frac{1}{2}(n+1)(n+2)$; 3. $\frac{1}{2^n n!}$.

Problem 2.9   "$\Rightarrow$" If $(a_n(x))_{n=0}^{\infty}$ is a Cauchy sequence then for every fixed $n \in \mathbb{N}$ there is some k_0 such that for $k, l \geq k_0$ the first n terms of $a_k(x)$ and $a_l(x)$ agree. Hence if we set $a_n := [x^n] a_k(x)$ for some $k \geq k_0$ then this definition will not depend on the choice of k. Now set $a(x) := \sum_{n=0}^{\infty} a_n x^n$ with a_n defined in this way. Then we have $\operatorname{ord}(a(x) - a_k(x)) > n$ for all $k > k_0$, hence $(a_n(x))_{n=0}^{\infty}$ converges to $a(x)$.
"$\Leftarrow$" Suppose that $(a_n(x))_{n=0}^{\infty}$ converges to $a(x)$. Let $n \in \mathbb{N}$ and $k_0 \in \mathbb{N}$ be such that for all $k \geq k_0$ we have $\operatorname{ord}(a(x) - a_k(x)) > n$. Then for $k, l \geq k_0$ we have $\operatorname{ord}(a(x) - a_k(x)) > n$ and $\operatorname{ord}(a(x) - a_l(x)) > n$. Now $\operatorname{ord}(a_k(x) - a_l(x)) = \operatorname{ord}((a(x) - a_l(x)) - (a(x) - a_k(x))) \geq \max(\operatorname{ord}(a(x) - a_l(x)), \operatorname{ord}(a(x) - a_k(x))) > n$, as desired.

Problem 2.10   Let $(K_n)_{n=0}^{\infty}$ be the sequence in question. We show by induction that $\operatorname{ord}(K_{n+1} - K_n) > n$ for all $n \in \mathbb{N}$. The claim then follows.
For $n = 0$ the claim is true because $\operatorname{ord}(\frac{x}{1-x} - x) = 2$. Assume now that it is true for some $n \in \mathbb{N}$. Then

$$
\operatorname{ord}(K_{n+1} - K_n) > n \Longrightarrow \operatorname{ord}((1 - K_{n+1}) - (1 - K_n)) > n
$$

$$
\Longrightarrow \operatorname{ord}\left(\frac{1}{1 - K_{n+1}} - \frac{1}{1 - K_n} \right) > n \Longrightarrow \operatorname{ord}\left(\frac{x}{1 - K_{n+1}} - \frac{x}{1 - K_n} \right) > n+1
$$

$$
\Longrightarrow \operatorname{ord}(K_{n+2} - K_{n+1}) > n+1,
$$

as desired.

Problem 2.12   Consider $B(x) := \frac{x}{\exp(x)-1} = \sum_{n=0}^{\infty} \frac{B_n}{n!} x^n$. We clearly have

$$
[x^n](\exp(x) - 1)B(x) = [x^n]x = 0 \qquad (n \geq 2).
$$

Now observe that

$$
(\exp(x) - 1)B(x) = \left(\sum_{n=1}^{\infty} \frac{1}{n!} x^n \right) \left(\sum_{n=0}^{\infty} \frac{B_n}{n!} x^n \right) = \sum_{n=0}^{\infty} \left(\sum_{k=0}^{n-1} \frac{1}{k!(n-k)!} B_k \right) x^n.
$$

Taking the coefficient of n and multiplying by $n!$ gives

$$[x^n](\exp(x)-1)B(x) = \sum_{k=0}^{n-1}\binom{n}{k}B_k \qquad (n\geq 0).$$

The claim follows.

Problem 2.14 From $D_x\exp(e^x-1) = e^x\exp(e^x-1)$ we obtain

$$\sum_{n=0}^{\infty}(n+1)\frac{B_{n+1}}{(n+1)!}x^n = \left(\sum_{n=0}^{\infty}\frac{1}{n!}x^n\right)\left(\sum_{n=0}^{\infty}\frac{B_n}{n!}x^n\right) = \sum_{n=0}^{\infty}\left(\sum_{k=0}^{n}\frac{1}{k!(n-k)!}B_k\right)x^n.$$

Comparing coefficients of x^n and multiplying by $n!$ on both sides gives the claim.

Problem 2.15 Writing $a(x) = \sum_{k=0}^{\infty}a_kx^k$, we have

$$[x^n]\frac{1}{1-x}a\left(\frac{x}{x-1}\right) = [x^n]\sum_{k=0}^{\infty}a_k(-1)^k\frac{x^k}{(1-x)^{k+1}} = \sum_{k=0}^{\infty}a_k(-1)^k\binom{n}{k}.$$

Now use that $\binom{n}{k} = 0$ for $k > n$.

Problem 2.16 The product rule implies $D_xb(x)^n = nb(x)^{n-1}b'(x)$ for all $n\geq 0$. Therefore, if $a(x) = \sum_{n=0}^{\infty}a_nx^n$ then

$$D_xa(b(x)) = \sum_{n=0}^{\infty}a_nD_xb(x)^n = \sum_{n=0}^{\infty}a_nnb(x)^{n-1}b'(x) = a'(b(x))b'(x),$$

the first step being justified by reference to the hint.

Problem 2.17 $1+\frac{1}{2}x^2-\frac{1}{8}x^4+\frac{1}{16}x^6-\frac{5}{128}x^8+\frac{7}{256}x^{10}+\cdots$

Problem 2.19 First set $q_n^{(1)} := 2q_{2n}-q_n$ to get rid of the term α_1/n. Next, because of $q_n^{(1)} = \beta/n^2+\cdots$ we have $q_{2n}^{(1)} = \beta/4n^2+\cdots$, so the quadratic term can be eliminated by taking $4q_{2n}^{(1)}-q_n^{(1)}$. Since this converges to three times the original limit, we set

$$q_n^{(2)} := \frac{1}{3}\left(4q_{2n}^{(1)}-q_n^{(1)}\right) = \frac{1}{3}(8q_{4n}-6q_{2n}+q_n).$$

For $q_n = C_{n+1}/C_n$ and $n=7$ this yields the estimate $\frac{719}{180}\approx 3.99444$ for the limit. The general formula for eliminating the first k terms in the asymptotic expansion is $q_n^{(k)} := (2^kq_{2n}^{(k-1)}-q_n^{(k-1)})/(2^k-1)$.

Problem 2.20 1. $(1-x-x^2)f(x) = x$; 2. $f(x)^2 = 1-x-x^2$; 3. $(1-x)^2f'(x)-(2x^2-4x+1)f(x) = 0$.

Problem 3.1 Yes. The high order coefficients of a truncated power series are considered unknown whereas the high order coefficients of a polynomial are zero. For example, if $a(x) = 1+x+x^2+O(x^3)$ is a truncated power series then $a(x)^2 = 1+$

$2x + 3x^2 + O(x^3)$ and we know nothing about the coefficient of x^3 in $a(x)^2$. On the other hand, if $a(x) = 1 + x + x^2$ is a polynomial, then $a(x)^2 = 1 + 2x + 3x^2 + 2x^3 + x^4$.

Problem 3.2 Yes. The object $\sum_{n=0}^{\infty}(1 + x^n)y^n$ belongs to $\mathbb{K}[x][[y]]$ but not to $\mathbb{K}[[y]][x]$.

Problem 3.3 Since we assume throughout the book that $\mathbb{K}$ is a field of characteristic zero, $\mathbb{K}[x]$ and $\mathrm{Pol}(\mathbb{K})$ are isomorphic. For fields of positive characteristic, this is not the case.

Problem 3.5 $x^{\overline{n}} = \sum_{k=0}^{n}(-1)^{n+k}S_1(n,k)x^k$; $x^n = \sum_{k=0}^{n}(-1)^{n+k}S_2(n,k)x^{\overline{k}}$.

Problem 3.6 1. $x^3 + 5x^2 + 5x^1 + 1$; 2. $2x^3 + 11x^2 + 3x^1 + 2$; 3. $5x^3 + 17x^2 + 7x^1 + 3$.

Problem 3.8 Let $a_k(x) := \sum_{n=k}^{\infty}\frac{S_2(n,k)}{n!}x^n$. We start with

$$S_2(n+1,k+1) = (k+1)S_2(n,k+1) + S_2(n,k) \qquad (n,k \geq 0).$$

Multiplying both sides by $x^n/n!$ and summing from k to ∞ gives

$$D_x a_{k+1}(x) = (k+1)a_{k+1}(x) + a_k(x) \qquad (k \geq 0).$$

For $k = 0$ we have $S_2(n,k) = \delta_{n,0}$, so $a_0(x) = 1$. Furthermore, we have $a_k(0) = 0$ for all $k > 0$. Together with this initial conditions, the differential equation determines each $a_k(x)$ uniquely. For $k = 1,2,3,\ldots$ we find

$$a_1(x) = e^x - 1, \quad a_2(x) = \tfrac{1}{2}(e^x - 1)^2, \quad a_3(x) = \tfrac{1}{6}(e^x - 1)^3, \ldots$$

and the general form $a_k(x) = \frac{1}{k!}(e^x - 1)^k$ is easily confirmed by induction on k.

For the result about the Bell numbers, observe that

$$\sum_{n=0}^{\infty}\frac{B_n}{n!}x^n = e^{e^x - 1} = \sum_{k=0}^{\infty}\frac{1}{k!}(e^x - 1)^k = \sum_{k=0}^{\infty}\sum_{n=k}^{\infty}\frac{S_2(n,k)}{n!}x^n = \sum_{n=0}^{\infty}\sum_{k=0}^{n}\frac{S_2(n,k)}{n!}x^n,$$

where in the last step we exploited that $S_2(n,k) = 0$ for $k < 0$ or $k > n$. The desired identity now follows by comparing coefficients and multiplying by $n!$.

Problem 3.9 1. The first identity is clear by $\exp(kx) = \sum_{d=0}^{\infty}k^d\frac{y^d}{d!}$. Next, we have

$$\sum_{k=0}^{n}\exp(ky) = \sum_{k=0}^{n}\exp(y)^k = \frac{\exp(y)^{n+1} - 1}{\exp(y) - 1}.$$

Multiplying by $y/(d+1)!$ and taking the coefficient of y^{d+1} on both sides leads to the desired identity.

2. The first identity is obtained from

$$\sum_{n=0}^{\infty}\frac{B_n(x)}{n!}y^n = \frac{y\exp(xy)}{\exp(y) - 1} = \frac{y}{\exp(y) - 1}\sum_{n=0}^{\infty}\frac{x^n}{n!}y^k = \left(\sum_{n=0}^{\infty}\frac{B_n}{n!}y^n\right)\left(\sum_{n=0}^{\infty}\frac{x^n}{n!}y^k\right)$$

$$= \sum_{n=0}^{\infty}\left(\sum_{k=0}^{n}\frac{B_k x^{n-k}}{k!(n-k)!}\right)y^k$$

by comparing coefficients of y^k and multiplying by $n!$. As a consequence, we get

$$B_{d+1}(n+1) - B_{d+1} = \sum_{k=0}^{d+1} B_k \binom{d+1}{k} (n+1)^{d+1-k} - B_{d+1}$$

$$= \sum_{k=0}^{d} B_k \binom{d+1}{k} (n+1)^{d-k+1}.$$

Problem 3.10   If $(H_n)_{n=0}^{\infty}$ were a polynomial sequence, then also $\Delta(H_n)_{n=0}^{\infty} = (\frac{1}{n+1})_{n=0}^{\infty}$.

Problem 3.11   In Mathematica:

```
genfun[poly_,x_] := Module[{c,n,k},
   c = Table[(−1)^{n+k} StirlingS2[n,k], {k,0,Length[c] − 1}, {n,0,Length[c] − 1}];
   c = c.CoefficientList[poly /. x → x − 1,x];
   Together[Sum[c[[k]](k − 1)!/(1 − x)^k, {k,1,Length[c]}]]]
```

Problem 3.13   1. $2x^4 + 10x^3 + 16x^2 + 11x + 6$; 2. no solution; 3. $x - 1$.

Problem 3.14   A similar reasoning as for the recurrence case gives the following case distinction: If $\deg q(x) + 1 \neq \deg r(x)$ then

$$\deg a(x) = \deg p(x) - \max(\deg q(x), \deg r(x)).$$

Otherwise, if $\operatorname{lc} r(x)/\operatorname{lc} q(x)$ is not an integer then $\deg a(x) = \deg p(x) - \deg r(x)$. Otherwise $\deg a(x) \leq \max(\deg p(x) - \deg r(x), \operatorname{lc} r(x)/\operatorname{lc} q(x))$.

Problem 3.15   A general bound is given in Sect. 7.4.

Problem 3.16   The chromatic polynomial is the same as for the graph in Fig. 3.3: $k^4 - 5k^3 + 8k^2 - 4k$. Setting $k = 1000$ yields 995007996000 colorings.

Problem 4.1   Use repeated squaring to compute $\phi^{2^{1000}}$ with a decent approximation of ϕ (a few hundred digits, say). Depending on the computer algebra system, it might be necessary to divide by some power of 10 from time to time. The result is $F_{2^{1000}} = 419087604\ldots$.

Problem 4.2   Use the logarithmic computation scheme for Fibonacci numbers and keep intermediate results reduced modulo 10^{10}, i.e., do the computations in the residue class ring $\mathbb{Z}_{10^{10}}$. The result is $F_{2^{1000}} = \ldots 48059253307$.

Problem 4.3 $F_{2n} + F_{2n+1} - F_{n+1}$. More generally, $\sum_{k=0}^{n} F_{m+k} = F_{n+m} + F_{n+m+1} - F_{m+1}$ for every $m \in \mathbb{N}$.

Problem 4.6   The finite continued fraction equals F_{n+2}/F_{n+1}. The infinite one is therefore ϕ.

Problem 4.8   If $p(x)$ is the characteristic polynomial of a C-finite recurrence satisfied by $(a_n)_{n=0}^{\infty}$, then letting this polynomial act as an operator on both sides of the inhomogeneous equation shows that $(u_n)_{n=0}^{\infty}$ satisfies the C-finite recurrence whose characteristic polynomial is $p(x)(x^r + c_{r-1}x^{r-1} + \cdots + c_1 x + c_0)$.

Problem 4.9 If $(H_n)_{n=0}^\infty$ is C-finite, then by Theorem 4.2 also $(H_{n+1} - H_n)_{n=0}^\infty$ is C-finite, so it suffices to show that $(1/(n+1))_{n=0}^\infty$ is not C-finite. Assume it were. Then

$$c_0 \frac{1}{n} + c_1 \frac{1}{n+1} + \cdots + c_r \frac{1}{n+r} = 0 \qquad (n \geq 1)$$

for some constants $c_0, \ldots, c_r$ with $c_r \neq 0$. Multiply by $n(n+1) \cdots (n+r)$ to obtain a polynomial relation of degree at most $r-1$. This relation, being valid for all integers $n \geq 1$, actually holds for all $n \in \mathbb{C}$. Setting $n = -r$ implies $c_r = 0$, a contradiction.

Problem 4.10 Induction on k. For $k = 1$ we have $S_2(n,1) = 1$ ($n \geq 0$), which is clearly C-finite. If $(S_2(n,k))_{n=0}^\infty$ is C-finite for some k, then the recurrence

$$S_2(n+1, k+1) - (k+1)S_2(n, k+1) = S_2(n,k) \qquad (n, k \geq 0),$$

in combination with the result of Problem 4.8, implies that $(S_2(n, k+1))_{n=0}^\infty$ is C-finite.

Problem 4.11 1. The identities are immediate consequences of Cassini's identity. It follows that $u(F_n, F_{n+1})^2 - 1 = 0$ for all $n \in \mathbb{N}$, and consequently

$$a(F_n, F_{n+1})(u(F_n, F_{n+1})^2 - 1) = 0 \qquad (n \in \mathbb{N})$$

for any $a(x,y) \in \mathbb{Q}[x,y]$.

2. Write $p(x,y) = p_0(x) + p_1(x)y + \cdots + p_d(x)y^d$ for $p_i(x) \in \mathbb{Q}[x]$. Proceed by induction on d. For $d < 2$ there is nothing to show. For $d \geq 2$ consider

$$\bar{p}(x,y) = p(x,y) - p_d(x)y^{d-2}(u(x,y) - 1).$$

We have $\deg_y \bar{p}(x,y) < 2$, and by the induction hypothesis there is some $\bar{a}(x,y)$ with $\bar{p}(x,y) = q(x,y) + \bar{a}(x,y)(u(x,y) - 1)$ and $q(x,y)$ at most linear in y. Setting $a(x,y) = \bar{a}(x,y) - p_d(x)y^{d-2}$ completes the induction step. The conclusion follows because $u(F_{2n}, F_{2n+1}) - 1 = 0$ for all $n \in \mathbb{N}$.

3. If at least one of $q_0(x)$ and $q_1(x)$ is nonzero then $d := \max(\deg q_0(x), 1 + \deg q_1(x))$ is a nonnegative integer. In this case, we have

$$\lim_{n \to \infty} \frac{q_0(F_{2n}) + q_1(F_{2n})F_{2n+1}}{F_{2n}^d} = [x^d]q_0(x) + \phi[x^{d-1}]q_1(x) \neq 0,$$

so we cannot have $q_0(F_{2n}) + q_1(F_{2n})F_{2n+1} = 0$ for all $n \in \mathbb{N}$ then. The conclusion follows by combining this result with the previous step.

4. The argument for (F_{2n+1}, F_{2n}) in place of (F_{2n}, F_{2n+1}) is fully analogous. (In the asymptotics argument, divide by F_{2n+1}^d and find $1/\phi$ instead of ϕ in the limit expression.) To get the final conclusion, observe that if $p(F_n, F_{n+1}) = 0$ for all $n \in \mathbb{N}$ then in particular for all even n, implying that $p(x,y)$ is a multiple of $u(x,y) - 1$, as well as for all odd n, implying that $p(x,y)$ is a multiple of $u(x,y) + 1$. Putting things together, $p(x,y)$ must be a multiple of $u(x,y)^2 - 1$, as claimed.

In general, if $(a_n^{(1)})_{n=0}^\infty, \ldots, (a_n^{(m)})_{n=0}^\infty$ are some sequences in a field $\mathbb{K}$, the set of all polynomials $p \in \mathbb{K}[x_1, \ldots, x_m]$ with $p(a_n^{(1)}, \ldots, a_n^{(m)}) = 0$ for all $n \in \mathbb{N}$ forms an ideal in the ring $\mathbb{K}[x_1, \ldots, x_m]$. If the $(a_n^{(i)})_{n=0}^\infty$ are C-finite in $\mathbb{Q}$, then a basis of this ideal can be computed by an algorithm [32].

Problem 4.12   No. For example, the solution of the recurrence $a_{n+2} = a_n$ is not uniquely determined by requesting that $a_0 = a_2 = 1$, because these conditions are satisfied by the two distinct solutions $(1)_{n=0}^\infty$ and $((-1)^n)_{n=0}^\infty$.

Problem 4.13   Because of $D_x^k \bar{a}(x) = \sum_{n=0}^\infty a_{n+k} \frac{x^n}{n!}$ for every k, there is a one-to-one correspondence between differential equations with constant coefficients for $\bar{a}(x)$ and recurrence equations with constant coefficients for $(a_n)_{n=0}^\infty$.

Problem 4.14   The characteristic polynomial of the recurrence is also the characteristic polynomial of the matrix M. Its roots $u_1, \ldots, u_r$ are therefore the eigenvalues of M. As they are distinct, it follows that M is equivalent to a diagonal matrix with $u_1, \ldots, u_r$ on the diagonal. A direct calculation confirms that $(1, u_i, \ldots, u_i^{r-1})$ is an eigenvector for u_i, and hence the representation $M = TDT^{-1}$ is established.

If $(a_n)_{n=0}^\infty$ is a solution of the recurrence, then

$$(a_n, a_{n+1}, \ldots, a_{n+r-1}) = M^n(a_0, a_1, \ldots, a_{r-1}) \quad (n \in \mathbb{N}).$$

Because of $M^n = (TDT^{-1})^n = TD^nT^{-1}$, this implies that $(a_n)_{n=0}^\infty$ can be written as a linear combination of $(u_i^n)_{n=0}^\infty$ $(i = 1, \ldots, r)$. Conversely, every vector $(a_0, a_1, \ldots, a_{r-1})$ of initial values gives rise to a solution $(a_n)_{n=0}^\infty$. This implies Theorem 4.1 for the present situation.

Problem 5.1   For instance via $1 \Rightarrow 4 \Rightarrow 3 \Rightarrow 2 \Rightarrow 1$.

Problem 5.2   Write $p(x,y) = p_0(x) + p_1(x)y + p_2(x)y + \cdots + p_d(x)y^d$ for $p_i(x) \in \mathbb{K}[x]$. Suppose one of the $p_i(x)$ is not the zero polynomial. Then there would be some n_0 with $p_i(n_0) \neq 0$. Then the univariate polynomial $p(n_0, y) \in \mathbb{K}[y]$ would not be the zero polynomial, although $p(n_0, m) = 0$ for all $m \in \mathbb{N}$. Contradiction.

Problem 5.3 $(p(n)a^n)_{n=0}^\infty$ where $p(x) \in \mathbb{K}[x]$ and $a \in \mathbb{K} \setminus \{0\}$.

Problem 5.4 $\frac{6}{\sqrt{5\pi}} n^{-2} 2^{5n} 3^{6n} 5^{-5n}$.

Problem 5.5   Substitute $f(x) = \sum_{n=0}^\infty \frac{a^{\bar{n}} b^{\bar{n}}}{c^{\bar{n}} n!} x^n$, $f'(x) = \sum_{n=0}^\infty \frac{(a+1)^{\bar{n}}(b+1)^{\bar{n}}}{(c+1)^{\bar{n}} n!} x^n$, and $f''(x) =$
$\sum_{n=0}^\infty \frac{(a+2)^{\bar{n}}(b+2)^{\bar{n}}}{(c+2)^{\bar{n}} n!} x^n$ into the left hand side of the differential equation, and check that the coefficient of x^n simplifies to zero for every $n \in \mathbb{N}$.

Problem 5.6   1. After multiplying the equation by $(1-x)^a$, compare coefficients on both sides. The coefficient of x^0 is 1 on both sides. For $n > 0$, the coefficient of

x^n on the left is

$$[x^n](1-x)^a {}_2F_1\left({a,b \atop c}\Big|x\right) = [x^n]\sum_{k=0}^{\infty}\binom{a}{k}(-1)^k x^k \sum_{k=0}^{\infty}\frac{a^{\overline{k}}b^{\overline{k}}}{c^{\overline{k}}k!}x^k$$

$$= \sum_{k=0}^{n}\binom{a}{n-k}(-1)^{n-k}\frac{a^{\overline{k}}b^{\overline{k}}}{c^{\overline{k}}k!}$$

and on the right hand side we have

$$[x^n]{}_2F_1\left({a,c-b \atop c}\Big|\frac{x}{x-1}\right) = \sum_{k=0}^{n}\frac{a^{\overline{k}}(c-b)^{\overline{k}}}{c^{\overline{k}}k!}\underbrace{[x^n]\frac{x^k}{(x-1)^k}}_{=(-1)^k\binom{n-1}{k-1}}.$$

It is therefore enough to prove the summation identity

$$\sum_{k=0}^{n}(-1)^{n-k}\frac{a^{\overline{k}}b^{\overline{k}}}{c^{\overline{k}}k!}\binom{a}{n-k} = \sum_{k=0}^{n}(-1)^k\frac{a^{\overline{k}}(c-b)^{\overline{k}}}{c^{\overline{k}}k!}\binom{n-1}{k-1} \qquad (n\geq 1).$$

This can be done with Zeilberger's algorithm. It will find that both sides satisfy the recurrence equation

$$(a-b-n)nS_n + (a(b-c-n-1)+(n+1)(b+c+2n+1))S_{n+1}$$
$$- (n+2)(c+n+1)S_{n+2} = 0 \qquad (n\geq 1).$$

After checking the identity for $n=1,2$, it follows for all $n\geq 0$ by induction.

2. ${}_2F_1\left({a,b \atop c}\Big|x\right) = (1-x)^{-a}{}_2F_1\left({a,c-b \atop c}\Big|\frac{x}{x-1}\right) = (1-x)^{-a}{}_2F_1\left({c-b,a \atop c}\Big|\frac{x}{x-1}\right)$

$= (1-x)^{-a}\left(1-\frac{x}{x-1}\right)^{-(c-b)}{}_2F_1\left({c-b,c-a \atop c}\Big|x\right) = (1-x)^{c-a-b}{}_2F_1\left({c-b,c-a \atop c}\Big|x\right).$

3. Multiply the Euler transform by $(1-x)^{a+b-c}$ and compare coefficients of x^n. Then on the right hand side there is just $\frac{(c-a)^{\overline{n}}(c-b)^{\overline{n}}}{c^{\overline{n}}n!}$ while on the left hand side we get

$$(1-x)^{a+b-c}{}_2F_1\left({a,b \atop c}\Big|x\right) = \sum_{n=0}^{\infty}\binom{a+b-c}{n}(-1)^n x^n \sum_{n=0}^{\infty}\frac{a^{\overline{n}}b^{\overline{n}}}{c^{\overline{n}}n!}x^n$$

$$= \sum_{n=0}^{\infty}\left(\begin{array}{c}\displaystyle\sum_{k=0}^{n}\frac{a^{\overline{k}}b^{\overline{k}}}{c^{\overline{k}}k!}\underbrace{\binom{a+b-c}{n-k}(-1)^{n-k}}_{=\frac{(c-a-b)^{\overline{n}}(-n)^{\overline{k}}}{n!(1+a+b-c-n)^{\overline{k}}}}\\ \underbrace{\qquad\qquad\qquad\qquad}_{=\frac{(c-a-b)^{\overline{n}}}{n!}{}_3F_2\left({-n,a,b \atop c,1+a+b-c-n}\Big|1\right)}\end{array}\right)x^n.$$

The identity follows.

Problem 5.7   Compare coefficients of x^n on both sides.

Problem 5.8   1. For every i with $\gcd(u_1(x), u_2(x+i)) \neq 0$ there must exist a point $(\xi, \zeta) \in \bar{\mathbb{K}}^2$ where ξ is a root of $u_1(x)$ and ζ is a root of $u_2(x)$ and $\xi = \zeta - i$. Since $u_1(x), u_2(x)$ each have only finitely many roots, there can be only finitely many pairs (ξ, ζ). Therefore, there can also be only finitely many such i.

2. First, we have

$$\frac{p(x+1)}{p(x)} \frac{\bar{u}_1(x)}{\bar{u}_2(x)} = \frac{g(x)g(x-1)\cdots g(x-i+1)}{g(x-1)g(x-2)\cdots g(x-i)} \frac{\bar{u}_1(x)}{\bar{u}_2(x)} = \frac{g(x)\bar{u}_1(x)}{g(x-i)\bar{u}_2(x)} = \frac{u_1(x)}{u_2(x)}.$$

Secondly, the choice of $g(x)$ implies $\gcd(\bar{u}_1(x), \bar{u}_2(x+i)) = 1$. Finally, it is clear that the set of all j with $\gcd(\bar{u}_1(x), \bar{u}_2(x+j)) \neq 1$ is contained in the set of all j with $\gcd(\bar{u}_1(x), \bar{u}_2(x+j)) \neq 1$ because $\bar{u}_1(x) \mid u_1(x)$ and $\bar{u}_2(x) \mid u_2(x)$.

3. The desired $i \in \mathbb{N}$ are precisely the positive integer roots of the univariate polynomial $\operatorname{res}_x(p(x), q(x+t)) \in \mathbb{K}[t]$.

4./5. In Maple:

```
gosperForm := proc(u,x)
    local p,q,r,i,j,k,g;
    p := 1; q := numer(u); r := denom(u);
    j := max(select(is, {solve(resultant(q, subs(x = x+i,r),x),i)},integer));
    while j >= 0 do
        g := gcd(q, subs(x = x+j,r));
        q := q/g; r := r/subs(x = x-j,g);
        p := p * product(subs(x = x-k,g),k = 1..j);
        j := max(select(is, {solve(resultant(q, subs(x = x+i,r),x),i)},integer));
    od;
    return([p,q,subs(x = x-1,r)]);
end;
```

Problem 5.9   For $p(x), q(x), r(x) \in \mathbb{K}[x]$ with $\gcd(q(x), r(x+i)) = 1$ for all $i \in \mathbb{N} \setminus \{0\}$ consider the equation

$$p(x) = q(x)y(x+1) - r(x)y(x).$$

If there are two different solutions $y_1(x), y_2(x) \in \mathbb{K}[x]$, then their difference $y_h(x) := y_1(x) - y_2(x)$ satisfies $q(x)y_h(x+1) = r(x)y_h(x)$. In this case we have

$$\frac{p(x+1)}{p(x)} \frac{q(x)}{r(x+1)} = \frac{p(x+1)/(y_h(x+1)r(x+1))}{p(x)/(y_h(x)r(x))},$$

so the summand sequence $(a_n)_{n=0}^{\infty}$ in question is essentially $a_n = \frac{p(n)}{y_h(n)r(n)}$.

Writing the general solution of the Gosper equation as $y(x) = y_1(x) + cy_h(x)$ with c a constant, Gosper's algorithm returns $w(n)a_n = \frac{r(n)}{p(n)}(y_1(n) + cy_h(n))\frac{p(n)}{y_h(n)r(n)} = \frac{y_1(n)}{y_h(n)} + c$. The choice of a solution of the Gosper equation therefore corresponds to a choice of the additive constant in the solution of the telescoping equation.

When $(a_n)_{n=0}^{\infty}$ is hypergeometric but not rational, then there is no such choice because then $(a_n)_{n=0}^{\infty}$ and the constant sequence $(1)_{n=0}^{\infty}$ are not similar.

Problem 5.10 Let $u_n = \prod_{k=0}^{n-1}\left(-\frac{c_0(k)}{c_1(k)}\right)$. Then $(u_n)_{n=0}^{\infty}$ is hypergeometric and $u_n \neq 0$ for all $n \in \mathbb{N}$. Substituting $s_n = u_n \bar{s}_n$ into the equation and dividing on both sides by $-c_0(n)u_n$ yields the new equation

$$\bar{s}_{n+1} - \bar{s}_n = -\frac{a_n}{c_0(n)u_n}.$$

This is now a telescoping equation with a hypergeometric sequence on the right hand side, so Gosper's algorithm can be used for finding its hypergeometric solutions. Every solution $(\bar{s}_n)_{n=0}^{\infty}$ gives rise to a solution $(u_n \bar{s}_n)_{n=0}^{\infty}$ of the original equation.

Problem 5.11 1. $\frac{x+1+n}{x+1}\binom{n+x}{n}$; 2. $(2n+1)4^{-n}\binom{2n}{n}$; 3. $(n+1)(m(m^2 - 7m + 3)\sqrt{5} - (3m^3 - 7m^2 + 19m - 6))/6(2m^3\sqrt{5} + (m^4 + 5m^2 - 1))$; 4. $2 - \frac{n!}{(2n+1)!}$; 5. $\frac{2}{3} + \frac{4(n-1)}{3(n+2)}4^n$; 6. $16^{-n}\binom{2n}{n}^2$.

Problem 5.12 Applying Gosper's algorithm to $a_k = \frac{1}{k}$ leads to the Gosper equation $1 = xy(x+1) - xy(x)$ which has obviously no polynomial solution $y(x)$.

Problem 5.13 Like in Zeilberger's algorithm, apply Gosper's algorithm to $(c_0 + c_1 k + \cdots + c_d k^d)/k!$ for a priori undetermined $c_0, \ldots, c_d$ and find suitable values for the c_i during the computation. The smallest d where a nontrivial solution can be found is $d = 1$. Here we get $c_0 = 1, c_1 = -1$, thus $p(x) = x - 1$.

Problem 5.14 1. $(2n-1)4^{n-1}$ $(n \geq 1)$; 2. $(2n+1)(-1)^n$ $(n \geq 0)$; 3. $2^{-n}\binom{2n}{n}$ $(n \geq 0)$; 4. 2^n $(n \geq 0)$; 5. $4^{-n}\binom{2n}{n}$ $(n \geq 0)$; 6. 0 $(n \geq 1)$.

Problem 5.15 $(n+2)^3 s_{n+2} - (2n+3)(17n^2 + 51n + 39)s_{n+1} + (n+1)^3 s_n = 0$ $(n \geq 0)$.

Problem 5.16 With $c_0(t) = 2t + 1$ and $c_1(t) = -2(t+1)$ we have

$$\int \left(c_0(n)\cos(\phi)^{2n} + c_1(n)\cos(\phi)^{2n+2}\right)d\phi = -\sin(\phi)\cos(\phi)^{2n+1}.$$

For the specific boundaries 0 and $\pi/2$ the right hand side becomes zero. Therefore the definite integral $I(n)$ on the right hand side of Wallis identity satisfies the recurrence equation $c_0(n)I(n) + c_1(n)I(n+1) = 0$ $(n \geq 0)$. Since the right hand side satisfies the same recurrence and both sides trivially agree for $n = 0$, the identity follows by induction.

Problem 6.1 $\frac{1+\sqrt{1-4x}}{2x} = \frac{1}{x} - 1 - x - 2x^2 + \cdots$ is not a power series.

Problem 6.2 $(4x^2 - x)a''(x) + (10x - 2)a'(x) + 2a(x) = 0$.

Problem 6.3 The recursive structure of the continued fraction implies that $K(x) = x/(1 - K(x))$. Clearing denominators gives the algebraic equation $K(x)^2 - K(x) + x = 0$. This equation has a unique formal power series solution whose constant term is zero. Since $xC(x)$ satisfies the same equation, we must have $K(x) = xC(x)$.

Problem 6.4 The vector space $V = \mathbb{K}(x) \oplus a(x)\mathbb{K}(x) \oplus \cdots \oplus a(x)^{d-1}\mathbb{K}(x)$ contains 1 and $a(x), a'(x), a''(x), \ldots$. Because of $\dim V = d$, any d elements are linearly dependent over $\mathbb{K}(x)$. In particular, $1, a(x), a'(x), \ldots, a^{(d-1)}(x)$ are linearly dependent. The dependence gives the desired equation.

Problem 6.5 $1 + \frac{4}{9}x + \frac{8}{243}x^2 + \frac{64}{6561}x^3 + \cdots;\quad -\frac{1}{2} - 4\sqrt{3}x^{1/2} - \frac{2}{9}x + \frac{2}{27\sqrt{3}}x^{3/2} + \cdots;$
$-\frac{1}{2} + 4\sqrt{3}x^{1/2} - \frac{2}{9}x - \frac{2}{27\sqrt{3}}x^{3/2} + \cdots.$

Problem 6.6 Observe that $\sum_{n=0}^{\infty} c_n x^n = a(x^2) + xb(x^2)$ and resort to Theorem 6.1.

Problem 6.7 $(x^2+x-1)^2 xy^2 + (x^2-x+1)(x^2+x-1)y + (x^4+2x^3-2x^2-x+1).$

Problem 6.8 $\frac{1}{2}\sqrt{5/6\pi}(4/5)^n n^{-3/2}.$

Problem 6.9 $4y^3 - (24x+3)y + 8x^2 + 20x - 1$ can be discovered by automated guessing as explained in Sect. 2.6. To prove that this "guessed" equation is correct, plug the series into the polynomial and simplify to zero (using summation algorithms whenever appropriate). Alternatively, convert the minimal polynomial into a differential equation (via Theorem 6.1), check compatibility of the series with this equation, and compare a suitable number of initial terms.

$\sum_{n=0}^{\infty} \binom{1/2}{n}^2 x^n$ is hypergeometric but not algebraic. $1/(1-x-x^2)$ is algebraic but not hypergeometric.

Problem 6.10 1. $a_{n,k} = \sum_j (-2)^{n-j}\binom{k+j}{j}\binom{k}{n-j}$ $(n,k \geq 0)$.

2. The recurrence is $a_{n+2,n+2} = -4\frac{n+1}{n+2}a_n$ $(n \geq 0)$. Together with the initial values $a_{0,0} = 1$ and $a_{1,1} = 0$ implies $a_{2n,2n} = (-1)^n\binom{2n}{n}$ and $a_{2n+1,2n+1} = 0$ $(n \geq 0)$.

3. $(4x^2+1)y^2 - 1 = 0$.

4. Plug $\sum_{n=0}^{\infty}\binom{2n}{n}(-x^2)^n$ into the left hand side of the equation and simplify the resulting expression to zero.

A proof for the general statement can be found in [54, Theorem 6.3.3].

Problem 6.12 Suppose $m(x,y) \in \mathbb{K}[x,y]$ is such that $m(x,a(x)) = 0$. Substituting $x \mapsto b(x)$ gives $m(b(x),a(b(x))) = 0$, so $m(b(x),x) = 0$. The claim follows.

Problem 6.13 1. We have $1/n! \sim cn^{-n+\frac{1}{2}}e^n$ for some constant c, whereas the coefficient sequences of algebraic power series grow like $n^{\alpha}d^n$ for some constants α, d.

2. Suppose $p(x,y) = p_0(x) + p_1(x)y + \cdots + p_d(x)y^d$ is an annihilating polynomial of $\exp(x)$. We may assume that d is minimal and that among all annihilating polynomials of degree d, the choice is made such that $\deg_x p_d(x)$ is as small as can be. Then $d \geq 1$ and $p_0(x)$ is not the zero polynomial. Differentiating $p(x,\exp(x)) = 0$ with respect to x implies that

$$q(x,y) := p_0'(x) + (p_1'(x) + p_1(x))y + \cdots + (p_d'(x) + dp_d(x))y^d$$

is another annihilating polynomial for $\exp(x)$. Now consider $q(x,y) - dp(x,y)$. This cannot be the zero polynomial because $d \neq 0$ and $\deg_x p_0'(x) < \deg_x p_0(x)$ implies that $[y^0](q(x,y) - dp(x,y)) = p_0'(x) - dp_0(x)$ is not the zero polynomial. But

$[y^d](q(x,y) - dp(x,y)) = p'_d(x)$. This is in contradiction to the minimality assumptions because either $p'_d(x) = 0$, then d was not minimal, or otherwise $\deg_x p'_d(x) < \deg_x p_d(x)$ and the degree of $p_d(x)$ was not minimal.

3. Suppose $p(x,y) = p_0(x) + p_1(x)y + \cdots + p_d(x)y^d$ is an annihilating polynomial of $\exp(x)$. We may assume that $p_d(x)$ is not the zero polynomial. Set $u = -\deg_x p_d(x)$ and $v = -d$. Then $\lim_{z \to \infty} p(z, e^z)z^u e^{vz} = \operatorname{lc} p_d(x) \neq 0$ while $p(z, e^z) = 0$ for all $z \in \mathbb{R}$. Contradiction.

4. If $\exp(x)$ is algebraic then so are $\exp(ix)$ and $\exp(-ix)$. Hence, also $\sin(x) = \frac{1}{2i}(\exp(ix) - \exp(-ix))$, and hence $x/\sin(x)$. But the latter power series, regarded as an analytic function, has a pole at $k\pi$ for every $k \in \mathbb{Z} \setminus \{0\}$. These are infinitely many. But the singularities of an algebraic function are roots of a univariate polynomial and therefore there are at most finitely many of them. Therefore, $x/\sin(x)$ cannot be algebraic, and therefore $\exp(x)$ cannot be algebraic either.

5. Suppose there is a nontrivial relation of the proposed form. Then $p_0(x)$ cannot be the zero polynomial, for otherwise multiplying by $n!$ on both sides would yield a contradiction to the linear independence statement of Theorem 4.1. Assume again that d is minimal and derive a smaller relation subtracting $dp_d(n+1)$ times the original relation from $(n+1)p_d(n)$ times the relation obtained from the original relation by shifting $n \mapsto n+1$. The first term in the resulting relation is the $dp_0(n)p_d(n+1) - (n+1)p_0(n+1)p_d(n)$ which cannot be identically zero because the $p_0(x)$ is not the zero polynomial and the two terms have different degree in n. This is the desired contradiction. To see that $\exp(x)$ is not algebraic, observe that comparing coefficients of x^n in a relation $p(x, \exp(x)) = 0$ would give rise to a nontrivial relation of the form we just proved to be impossible.

Problem 6.14 If $\log(1+x)$ was algebraic, then, since $\exp(\log(1+x)) = 1+x$, also $\exp(x)$ would be algebraic by Problem 6.12. This is not the case by Problem 6.13, so $\log(1+x)$ is not algebraic either. Now $\sum_{n=0}^{\infty} H_n x^n = -\frac{\log(1-x)}{1-x}$ cannot be algebraic either, for if it was, then multiplying by $x-1$ and substituting $x \mapsto -x$ would yield again an algebraic series, while $\log(1+x)$ was shown to be not algebraic.

Problem 6.15 1. Counterexample: $\sum_{n=0}^{\infty} 2^n x^n = 1/(1-2x)$ is algebraic as series, but 2^n grows too quickly to be algebraic as a sequence.

2. Counterexample: $a_n = 1/(n+1)$ is algebraic as a sequence, because $(n+1)a_n - 1 = 0 \ (n \geq 0)$, but $\sum_{n=0}^{\infty} a_n x^n = \log(1-x)$ is not algebraic as a series.

Problem 6.16 C_n.

Problem 6.17 $\frac{1-\sqrt{1-4x}}{2x\sqrt{1-4x}} = 1 + 3x + 10x^2 + 35x^3 + \cdots$.

Problem 7.2 1. $\frac{1}{4}n((1-n) + 2(n+1)H_n)$; 2. $2n - (2n+1)H_n + (n+1)H_n^2$; 3. $(n+1)H_n^{(2)} - H_n$.

Problem 7.3 1. n; 2. 2^n; 3. $2^n + 3^n$; 4. $n!$; 5. $\binom{2n}{n}$; 6. $2^n + \binom{2n}{n}$; 7. $n!$; 8. $n! + (2n)!$; 9. $1/(n! + (2n)!)$.

Problem 7.4   Let $c_n = b_n - a_n$. Then $c_n \neq 0$ for finitely many $n \in \mathbb{N}$ only. Therefore $c(x) := \sum_{n=0}^{\infty} c_n x^n$ is a polynomial, and thus holonomic. (It satisfies, for instance, the differential equation $c'(x)a(x) - c(x)a'(x) = 0$.) Since $(a_n)_{n=0}^{\infty}$ is holonomic and also $(c_n)_{n=0}^{\infty}$ and $b_n = a_n + c_n$ ($n \in \mathbb{N}$), it follows that $(b_n)_{n=0}^{\infty}$ is holonomic.

The corresponding statement is true for C-finite sequences and for coefficient sequences of algebraic power series, but not for hypergeometric sequences.

Problem 7.7   All.

Problem 7.8   Let $a(x)$ be the series in question. Then for every $k \in \mathbb{N}$ there is some polynomial $q_k(x) \in \mathbb{Q}[x]$ of degree k such that $D_x^k a(x) = q_k(\exp(x))a(x)$. This follows directly from a repeated application of the chain rule. Now if there was an equation

$$p_0(x)a(x) + p_1(x)a'(x) + \cdots + p_r(x)D_x^r a(x) = 0$$

then this would imply

$$\big(p_0(x) + p_1(x)q_1(\exp(x)) + \cdots + p_r(x)q_r(\exp(x))\big)a(x) = 0.$$

Dividing by $a(x)$ gives a polynomial equation for $\exp(x)$ which is nontrivial because $\deg q_k(x) = k$ for every k. We have reached a contradiction to the result of Problem 6.13 and therefore a holonomic differential equation for $a(x)$ cannot exist.

Problem 7.9 $a_k := (\frac{1}{2})^k \binom{2k}{k}$ is clearly holonomic as a sequence. Hence $a(x) := \sum_{k=0}^{\infty} a_k x^k$ is holonomic as a power series. Hence $\frac{1}{1-x}a(\frac{x}{x-1})$ is holonomic as a power series. Hence $[x^n]\frac{1}{1-x}a(\frac{x}{x-1})$ is holonomic as a sequence. By the Euler transform, the latter is equal to s_n.

Problem 7.10   Starting from the obvious equations satisfied by $1/\sqrt{1-4t^2}$, $\exp(t)$, and $4t(xy - t(x^2 + y^2))/(1 - 4t^2)$, construct a differential equation for the right hand side. This gives

$$(1 - 4t^2)^2 a'(t) + 4((2x^2 + 2y^2 - 1)t - 4xyt^2 - xy + 4t^3)a(t) = 0.$$

This differential equation translates into the recurrence equation

$$16(n+1)a_n - 16xya_{n+1} - 4(2n - 2x^2 - 2y^2 + 5)a_{n+2} - 4xya_{n+3} + (n+4)a_{n+4} = 0$$

valid for $n \geq 0$. Now verify that $H_n(x)H_n(y)/n!$ satisfies this recurrence as well and compare four initial values.

Problem 7.11   Start with the differential equation from Problem 5.5 and the obvious equations satisfied by $4x/(1+x)^2$, $(1+x)^{2a}$ and x^2, and construct differential equations for the series on the left hand side and the right hand side of the equation. It turns out that both sides satisfy

$$4ab(x-1)f(x) + 2(x+1)(-2ax + bx^2 + b - x^2)f'(x) - (x-1)x(x+1)^2 f''(x) = 0.$$

Finally, compare a suitable number of initial values.

Problem 7.12 The recurrence corresponding to the differential equation is

$$52a_n+(2705-104n)a_{n+1}+2(26n^2-1353n-54)a_{n+2}+53(n-50)(n+3)a_{n+3}=0.$$

It has a singularity at $n=50$, so fixing a_0,a_1,a_2 does not determine the value a_{53}.

Problem 7.13 See Sect. 8.3 of [44].

Problem 7.15 $h_n=n!$. To obtain the other solution, plug the proposed form of u_n into the equation. Using that h_n is a solution, the equation can be simplified to a first order equation for $\bar{h}_n$. Its solution is $\bar{h}_n=(-1)^n/n!$.

More generally, whenever $(h_n)_{n=0}^\infty$ is some solution to a recurrence of order r, there is a second solution $(u_n)_{n=0}^\infty$ with $u_n=h_n\sum_{k=0}^{n-1}\bar{h}_k$ where $(\bar{h}_n)_{n=0}^\infty$ satisfies a recurrence of order $r-1$.

Problem 7.16 $\xi=1$ or $\xi=-1$.

Problem 7.17 1. If $(h_n)_{n=0}^\infty$ is such that $u(n)h_{n+1}+v(n)h_n$ ($n\geq0$) for some polynomials $u(x),v(x)\in\mathbb{K}[x]$, then every solution $(a_n)_{n=0}^\infty$ of the given inhomogeneous equation will also satisfy the homogeneous equation

$$v(n)p_0(n)a_n+(u(n)p_0(n+1)+v(n)p_1(n))a_{n+1}+\cdots$$
$$+(u(n)p_{r-1}(n+1)+v(n)p_r(n))a_{n+r}+u(n)p_r(n+1)a_{n+r+1}=0 \quad (n\geq0).$$

Use Petkovšek's algorithm to determine the hypergeometric solutions of this equation, and then check which of these also satisfies the original equation. Return those as answer and discard the others.

2. Petkovšek's algorithm returns the shift quotients $w(x)\in\mathbb{K}(x)$ of all the hypergeometric solutions. We need to detect for a given $w(x)\in\mathbb{K}(x)$ whether it is actually the shift quotient of a rational function, i.e., whether there exists $u(x)\in\mathbb{K}(x)$ such that $u(x+1)/u(x)=w(x)$. This can be done by computing a Gosper form of $w(x)$: If $w(x)=p(x+1)q(x)/p(x)/r(x+1)$ is a Gosper form of $w(x)$, then $w(x)$ is the shift quotient of a rational function if and only if $q(x)=r(x)$. In this case $u(x)=p(x)/q(x)$.

It should be remarked that this is a rather brutal way of finding rational solutions of holonomic recurrence equations. Better algorithms can be found in the literature [2].

Problem 7.18 Induction on k. For $k=0$ we have $S_1(n,0)=0$ ($n\geq1$), which is clearly holonomic. If $(S_1(n,k))_{n=0}^\infty$ is holonomic for some $k\geq0$, then the general recurrence

$$S_1(n,k)+(n-1)S_1(n-1,k)=S_1(n-1,k-1) \quad (n,k>0)$$

for Stirling numbers of the first kind (Ex. 3.7) implies

$$S_1(n,k+1)=(-1)^n(n-1)!\sum_{i=0}^{n-1}\frac{S_1(i,k)}{(-1)^i i!} \quad (n>0),$$

so $(S_1(n,k+1))_{n=0}^\infty$ is holonomic as well.

The claimed harmonic number representations can be checked by explicitly computing recurrence equations for the $(S_1(n,k))_{n=0}^\infty$ $(k = 1,2,3,4)$, checking that the claimed expressions satisfy these recurrence equations, and comparing a suitable number of initial values.

Problem 7.19   1. The worst case situation is exactly the same as for Quicksort.

2. $c_{n,i} = (n-1) + \frac{1}{n}\left(\sum_{k=1}^{i-1} c_{n-k,i-k} + \sum_{k=i+1}^{n} c_{k-1,i}\right)$ and $c_{n,1} = c_{n,n} = n-1$.

Elimination of the summation signs leads to the recurrence

$$c_{n,i} - c_{n+1,i} - c_{n+1,i+1} + c_{n+2,i+1} = \frac{2}{n+2} \qquad (n \geq 1, 1 < i < n).$$

3. The terms for $n = 1,2,\ldots,30$ read $0, 1, \frac{8}{3}, \frac{53}{12}, \frac{197}{30}, \frac{87}{10}, \frac{467}{42}, \frac{1133}{84}, \frac{1013}{63}, \frac{23447}{1260}, \frac{49249}{2310}, \frac{664327}{27720}, \frac{4822549}{180180}, \frac{118115}{4004}, \frac{1166365}{36036}, \frac{1810561}{51480}, \frac{58338047}{1531530}, \frac{27874461}{680680}, \frac{511162633}{11639628}, \frac{5449841867}{116396280}, \frac{161087299}{3233230}, \frac{6141864151}{116396280}, \frac{74689129967}{1338557220}, \frac{10489933451}{178474296}, \frac{29559627277}{478056150}, \frac{867902775947}{13385572200}, \frac{1363691268007}{20078358300}, \frac{5698145622329}{80313433200}, \frac{5748829529089}{77636318760}, \frac{89786686813897}{1164544781400}.$

4. We found the equation

$$x(x+1)(x-1)^3(2x^6 + 2x^5 - 18x^4 - 17x^3 - 9x^2 - 30x - 30)a^{(3)}(x)$$
$$+ 2(x-1)^2(9x^8 + 15x^7 - 94x^6 - 148x^5 - 83x^4 - 207x^3 - 282x^2 + 90)a''(x)$$
$$+ 2(x-1)(15x^8 + 24x^7 - 175x^6 - 274x^5 - 320x^4 - 582x^3$$
$$- 528x^2 + 480x + 360)a'(x)$$
$$+ 4(3x^7 + 11x^6 - 40x^5 - 227x^4 - 195x^3 - 42x^2 + 300x + 90)a(x) = 0.$$

5. We found $c_1(x) = \frac{x^2 - 3}{(x-1)^2 x^4}$, $c_2(x) = \frac{1}{(x-1)^2}$, and

$$c_3(x) = \frac{x^6 - 18 + 2(x^2 - 3)\log(x+1) + 2(2x^4 + x^2 - 3)\log(1-x)}{(x-1)^2 x^4}.$$

6. For the above choice of $c_1(x), c_2(x), c_3(x)$ we get $\alpha_1 = 3, \alpha_2 = -\frac{1}{2}, \alpha_3 = -\frac{1}{2}$.

7. On $U := \{z \in \mathbb{C} : |z| < 1\}$ we can define the analytic function

$$c: U \to \mathbb{C}, \quad c(z) = \alpha_1 c_1(z) + \alpha_2 c_1(z) + \alpha_3 c_3(z).$$

It has singularities at $z = \pm 1$ and there we have

$$c(z) \sim \tfrac{1}{2}(1 + \log(1+z)) \quad (z \to -1)$$
$$c(z) \sim \frac{2(1 + \log(2))}{(1-z)^2} \quad (z \to 1).$$

The growth implied by the second estimate dominates the growth implied by the first. Therefore

$$[x^n]c(x) \sim 2(1 + \log(2))n \quad (n \to \infty).$$

8. $\frac{2n^5 + 28n^4 + 123n^3 + 166n^2 - 81n - 210}{(n+1)(n+2)(n+3)(n+4)} + \frac{n^4 + 15n^3 + 85n^2 + 209n + 198}{(n+1)(n+2)(n+3)(n+4)}(-1)^n - 2(n+6)\sum_{k=1}^{n}\frac{(-1)^k}{k} -$

$10H_n$.

A.7 Bibliographic Remarks

Chapter 1

Hoare's original paper on Quicksort is [29]; the complexity analysis given here already appears there and is used as an example for the analysis of algorithms in many introductory textbooks. Further examples for analysis of algorithms can be found in the seminal volumes of Knuth [34, 33]. The book Concrete Mathematics [24] arose as an offspring of Knuth's ground breaking work in this area.

Havil [27] gives a fine and very readable account on the mysteries surrounding Euler's magic constant γ.

Chapter 2

Wilf's introduction to generating functions [62] contains further information and additional examples for the usage of formal power series in combinatorics and other branches of mathematics. Ongoing research on formal power series is presented at the annual meetings of the international conference series FPSAC ("Formal Power Series and Algebraic Combinatorics").

Sokal [52] gives a "ridiculously simple" version of the implicit function theorem for analytic functions as well as for formal power series.

Sloane's collection of integer sequences has originally appeared as a book [51]. One of the first uses of computers for detecting possible equations among sequences or power series was by Pivar and Finkelstein [45], the idea has subsequently been adapted to various different types of equations [49, 39, 37, 31, 28].

Richardson's convergence acceleration technique dates back to 1910 [47], a survey of more recent developments can be found in [13].

Chapter 3

Most of the material in this chapter is part of the mathematical folklore.

We follow Concrete Mathematics for the notation of rising and falling factorials; many other notations are used in the literature. In particular, the symbol $(x)_n$ may refer to $x^{\overline{n}}$ as well as to $x^{\underline{n}}$.

For a collection of additional facts on Stirling numbers, see Sect. 6.1 of Concrete Mathematics. Also Stirling numbers appear in a variety of different notations.

The algorithm for solving first order inhomogeneous linear recurrence equations with polynomial coefficients appears in Gosper's paper [23].

The theory behind the examples for partition analysis was developed in the late 19th century by MacMahon [38] and revived and brought to the computer by Andrews, Paule and Riese in a series of articles. The examples given here are taken from [9]. For more recent work, see [8] and the references given there.

Chapter 4

The term C-finite was coined by Zeilberger in [65], he also calls holonomic recurrences *P-finite* because they have polynomial coefficients where C-finite recurrences have constant coefficients.

Stanley discusses rational generating functions in Chap. 4 of [53]. Combinatorial examples as well as pointers to the literature can be found there. For number theoretic aspects of C-finite sequences we refer to the comprehensive text [20].

A general summation algorithm for sums over C-finite sequences is due to Greene and Wilf [25]. Their algorithm includes as a special case the one we describe.

For general aspects of the theory of orthogonal polynomials, see Chap. 6 in the book of Temme [55] and the relevant chapters of [7].

Chapter 5

Andrews, Askey and Roy [7] is excellent reference for classical aspects of the theory of hypergeometric series.

Gosper's algorithm originally appeared in [23], Paule [41] provides an algebraic explanation of it. The original article of Zeilberger's algorithm is [64]. There are also textbooks solely devoted to hypergeometric summation [44, 35], and further references to the literature are given there.

The elliptic arc length example is taken from [6].

Monthly problems which can be solved with the help of summation algorithms are collected in [40]. Also the examples we have given are taken from there.

Chapter 6

The kernel method first appeared as a solution to Exercise 2.2.1-4 in [33], see [10, 12] and the references given there for generalizations.

The solution of algebraic equations in terms of Puiseux series was already proposed almost two centuries before Puiseux by Isaac Newton.

Buchberger's theory of Gröbner bases [17] provides a computational alternative to resultant computations.

More on the connection of context free languages to algebraic power series can be found in [21]. Also detailed background information on asymptotic techniques can be found there.

Facts about Legendre polynomials are collected in [18].

Chapter 7

Concrete Mathematics [24] discusses harmonic numbers and their summation in Sects. 6.3 and 6.4. For algebraic relations among generalized harmonic numbers, their automated computation, and their relevance in particle physics, we refer to [1] and the references give there. Apery's proof of the irrationality of $\zeta(3)$ can be found in [56].

Zeilberger promoted holonomic sequences and power series in [65], there he discusses a more general definition applicable to sequences and series in several variables. Closure properties can be proven and computed also in this case. Stanley treats holonomic objects in Sect. 6.4 of [54] under the names D-finite (for power series) and P-recursive (for sequences). Summation algorithms for general multivariate holonomic sequences and functions are described in [15] and [14].

The solution of linear differential equations in terms of generalized series goes back to the work of Frobenius at the end of the 19th century.

The criterion that $a(x)$ and $1/a(x)$ are holonomic if and only if $a'(x)/a(x)$ is algebraic can be found as Exercise 1.39 in [57]. The original source is [26].

Petkovšek's algorithm first appeared in [43] and is also described in [44, 35]. Abramov and van Hoeij present their summation algorithm in [4].

Proofs for the claimed permutation statistics are given in [62]. For techniques for high performance computations of constants, see the book [11]. Bessel functions are described at length in the classical book [60].

The fast median search algorithm discussed in Problem 7.19 is described in Chap. 9 of [16].

References

1. Jakob Ablinger. A computer algebra toolbox for harmonic sums related to particle physics. Master's thesis, J. Kepler University, Linz, February 2009.
2. Sergei A. Abramov. Rational solutions of linear difference and q-difference equations with polynomial coefficients. In: *Proceedings of ISSAC'95*, July 1995.
3. Sergei A. Abramov, Jacques J. Carette, Keith O. Geddes, and Ha Q. Le. Telescoping in the context of symbolic summation in Maple. *Journal of Symbolic Computation*, 38(4):1303–1326, 2004.
4. Sergei A. Abramov and Mark van Hoeij. Integration of solutions of linear functional equations. *Integral Transforms and Special Functions*, 8(1-2):3–12, 1999.
5. Milton Abramowitz and Irene A. Stegun. *Handbook of Mathematical Functions*. Dover Publications, Inc., 9th edn., 1972.
6. Gert Almkvist and Bruce Berndt. Gauss, Landen, Ramanujan, the arithmetic-geometric mean, ellipses, π, and the ladies diary. *The American Mathematical Monthly*, 95(7):585–608, 1988.
7. George E. Andrews, Richard Askey, and Ranjan Roy. *Special Functions*. Encyclopedia of Mathematics and its Applications, vol. 71. Cambridge University Press, 1999.
8. George E. Andrews, Peter Paule, and Axel Riese. MacMahon's partition analysis XI: Broken diamonds and modular forms. *Acta Arithmetica*, 126:281–294, 2007.
9. George E. Andrews, Peter Paule, Axel Riese, and Volker Strehl. MacMahon's partition analysis V: Bijections, recursions, and magic squares. In: *Algebraic Combinatorics and Applications*, pp. 1–39. Springer, 2001.
10. Cyril Banderier and Philippe Flajolet. Basic analytic combinatorics of directed lattice paths. *Theoretical Computer Science*, 281(1–2):37–80, 2002.
11. Lennart Berggren, Jonathan M. Borwein, and Peter Borwein. *Pi: A Source Book*. Springer, 3rd edn., 2004.
12. Mireille Bousquet-Mélou and Marko Petkovšek. Walks confined in a quadrant are not always D-finite. *Theoretical Computer Science*, 307(2):257–276, 2003.
13. Claude Brezinski. Convergence acceleration during the 20th century. *Journal of Computational and Applied Mathematics*, 122:1–21, 2000.
14. Frédéric Chyzak. An extension of Zeilberger's fast algorithm to general holonomic functions. *Discrete Mathematics*, 217:115–134, 2000.
15. Frédéric Chyzak and Bruno Salvy. Non-commutative elimination in Ore algebras proves multivariate identities. *Journal of Symbolic Computation*, 26:187–227, 1998.
16. Thomas H. Cormen, Charles E. Leiserson, Ronald L. Rivest, and Clifford Stein. *Introduction to Algorithms*. Cambridge, Mass., 2nd edn., 2001.
17. David Cox, John Little, and Donal O'Shea. *Ideals, Varieties, and Algorithms*. Springer, 1992.
18. T. Mark Dunster. *Legendre and Related Functions*, chap. 14. Cambridge University Press, 2010.

19. Frank W. J. Olver et al. (Ed.) *NIST Handbook of Mathematical Functions*. Cambridge, 2010.
20. Graham Everest, Alf van der Poorten, Igor Shparlinski, and Thomas Ward. *Recurrence Sequences*. Mathematical Surveys and Monographs, vol. 104. American Mathematical Society, 2003.
21. Philippe Flajolet and Robert Sedgewick. *Analytic Combinatorics*. Cambridge University Press, 2009.
22. Keith O. Geddes, Stephen R. Czapor, and George Labahn. *Algorithms for Computer Algebra*. Kluwer, 1992.
23. William Gosper. Decision procedure for indefinite hypergeometric summation. *Proceedings of the National Academy of Sciences of the United States of America*, 75:40–42, 1978.
24. Ronald L. Graham, Donald E. Knuth, and Oren Patashnik. *Concrete Mathematics*. Addison-Wesley, 2nd edn., 1994.
25. Curtis Greene and Herbert S. Wilf. Closed form summation of C-finite sequences. *Transactions of the American Mathematical Society*, 359:1161–1189, 2007.
26. William A. Harris and Yasutaka Sibuya. The reciprocals of solutions of linear ordinary differential equations. *Advances in Mathematics*, 58(2):119–132, 1985.
27. Julian Havil. *Gamma: exploring Euler's constant*. Princeton University Press, 2003.
28. Waldemar Hebisch and Martin Rubey. Extended Rate, more GFUN. ArXiv:0702086v2, 2010.
29. Charles A. R. Hoare. Quicksort. *Computer Journal*, 5(1):10–15, 1962.
30. Edward L. Ince. *Ordinary Differential Equations*. Longmans, Green and Co., 1926.
31. Manuel Kauers. Guessing handbook. Technical Report 09-07, RISC-Linz, 2009.
32. Manuel Kauers and Burkhard Zimmermann. Computing the algebraic relations of C-finite sequences and multisequences. *Journal of Symbolic Computation*, 43(11):765–844, 2008.
33. Donald E. Knuth. *The Art of Computer Programming*, vol. I. Fundamental Algorithms. Addison-Wesley, 3rd edn., 1997.
34. Donald E. Knuth. *The Art of Computer Programming*, vol. II. Seminumerical Algorithms. Addison-Wesley, 3rd edn., 1997.
35. Wolfram Koepf. *Hypergeometric Summation*. Vieweg, 1998.
36. Christoph Koutschan. *Advanced Applications of the Holonomic Systems Approach*. PhD thesis, RISC-Linz, Johannes Kepler Universität Linz, 2009.
37. Christian Krattenthaler. RATE: A Mathematica guessing machine. Available at http://mat.univie.ac.at/˜kratt/rate/rate.html, 1997.
38. Percy A. MacMahon. *Combinatory Analysis*. Cambridge University Press, 1915–1916.
39. Christian Mallinger. Algorithmic manipulations and transformations of univariate holonomic functions and sequences. Master's thesis, J. Kepler University, Linz, August 1996.
40. István Nemes, Marko Petkovšek, Herbert S. Wilf, and Doron Zeilberger. How to do monthly problems with your computer. *American Mathematical Monthly*, 104:505–519, 1997.
41. Peter Paule. Greatest factorial factorization and symbolic summation. *Journal of Symbolic Computation*, 20:235–268, 1995.
42. Peter Paule and Markus Schorn. A Mathematica version of Zeilberger's algorithm for proving binomial coefficient identities. *Journal of Symbolic Computation*, 20(5–6):673–698, 1995.
43. Marko Petkovšek. Hypergeometric solutions of linear recurrences with polynomial coefficients. *Journal of Symbolic Computation*, 14(2–3):243–264, 1992.
44. Marko Petkovšek, Herbert Wilf, and Doron Zeilberger. $A = B$. AK Peters, Ltd., 1997.
45. Malcom Pivar and Mark Finkelstein. Automation, using LISP, of inductive inference on sequences. In: *The Programming Language LISP: Its Operations and Applications*, pp. 125–136. Information International, Inc., 1964.
46. George Polya. *Mathematics and Plausible Reasoning: Induction and analogy in mathematics*. Princeton University Press, 1954.
47. Lewis Fry Richardson. The approximate arithmetic solution by finite difference of physical problems involving differential equations, with an application to the stress in a masnory dam. *Philosophical Transactions of the Royal Society London, Series A*, 210:307–357, 1910.
48. Theodore J. Rivlin. *Chebyshev polynomials: From approximation theory to algebra and number theory*. John Wiley, 2nd edn., 1990.

49. Bruno Salvy and Paul Zimmermann. Gfun: a Maple package for the manipulation of generating and holonomic functions in one variable. *ACM Transactions on Mathematical Software*, 20(2):163–177, 1994.
50. Carsten Schneider. Symbolic summation assists combinatorics. *Sem. Lothar. Combin.*, B56b, 2007.
51. Neil J. A. Sloane and Simon Plouffe. *The Encyclopedia of Integer Sequences*. Academic Press, 1995.
52. Alan D. Sokal. A ridiculously simple and explicit implicit function theorem. *Seminaire Lotharingien de Combinatoire*, 61A, 2009.
53. Richard P. Stanley. *Enumerative Combinatorics, vol. 1*. Cambridge Studies in Advanced Mathematics 49. Cambridge University Press, 1997.
54. Richard P. Stanley. *Enumerative Combinatorics, vol 2*. Cambridge Studies in Advanced Mathematics 62. Cambridge University Press, 1999.
55. Nico M. Temme. *Special Functions*. John Wiley & Sons, 1996.
56. Alfred van der Poorten. A proof that Euler missed... — Apéry's proof of the irrationality for $\zeta(3)$. *The Mathematical Intelligencer*, 1:195–203, 1979.
57. Marius van der Put and Michael F. Singer. *Galois Theory of Linear Differential Equations*. Springer, 2003.
58. Joachim von zur Gathen and Jürgen Gerhard. *Modern Computer Algebra*. Cambridge University Press, 1999.
59. Robert J. Walker. *Algebraic Curves*. Princeton University Press, 1950.
60. G. N. Watson. *A Treatise on the Theory of Bessel functions*. Cambridge University Press, 2nd edn., 1995.
61. Kurt Wegschaider. Computer generated proofs of binomial multi-sum identities. Master's thesis, RISC-Linz, May 1997.
62. Herbert S. Wilf. *generatingfunctionology*. AK Peters, Ltd., 1989.
63. Franz Winkler. *Polynomial Algorithms in Computer Algebra*. Springer, 1996.
64. Doron Zeilberger. A fast algorithm for proving terminating hypergeometric identities. *Discrete Mathematics*, 80:207–211, 1990.
65. Doron Zeilberger. A holonomic systems approach to special function identities. *Journal of Computational and Applied Mathematics*, 32:321–368, 1990.

Subject Index